1 MONTH OF FREE READING

at

www.ForgottenBooks.com

By purchasing this book you are eligible for one month membership to ForgottenBooks.com, giving you unlimited access to our entire collection of over 1,000,000 titles via our web site and mobile apps.

To claim your free month visit: www.forgottenbooks.com/free106958

* Offer is valid for 45 days from date of purchase. Terms and conditions apply.

ISBN 978-1-5282-7554-5
PIBN 10106958

This book is a reproduction of an important historical work. Forgotten Books uses state-of-the-art technology to digitally reconstruct the work, preserving the original format whilst repairing imperfections present in the aged copy. In rare cases, an imperfection in the original, such as a blemish or missing page, may be replicated in our edition. We do, however, repair the vast majority of imperfections successfully; any imperfections that remain are intentionally left to preserve the state of such historical works.

Forgotten Books is a registered trademark of FB &c Ltd.
Copyright © 2018 FB &c Ltd.
FB &c Ltd, Dalton House, 60 Windsor Avenue, London, SW19 2RR.
Company number 08720141. Registered in England and Wales.

For support please visit www.forgottenbooks.com

THE USE OF THE BLOWPIPE

IN CHEMICAL ANALYSIS,

AND IN

THE EXAMINATION OF MINERALS

BY J. J. BERZELIUS,

MEMBER OF THE ACADEMY OF SCIENCES OF STOCKHOLM, &c.

TRANSLATED FROM THE FRENCH OF M. FRESNEL,

BY J. G. CHILDREN,

FRS. L. & E., FLS., MGS., &c.

WITH A SKETCH OF BERZELIUS' SYSTEM OF MINERALOGY; A SYNOPTIC TABLE OF THE PRINCIPAL CHARACTERS OF THE PURE EARTHS AND METALLIC OXIDES BEFORE THE BLOWPIPE, AND NUMEROUS NOTES AND ADDITIONS BY THE TRANSLATOR.

LONDON:
PRINTED FOR BALDWIN, CRADOCK, AND JOY,
PATERNOSTER-ROW;

TRANSLATOR'S PREFACE.

I FEEL confident that no apology for presenting this translation to the public, nor any eulogy on the author of the original work, are at all necessary. The name of Berzelius, as a skilful and patient experimenter, stands almost unrivalled; and the present Essay amply vindicates his claim to the high reputation he has acquired. It is an invaluable collection of important and new facts, and admirably supplies the want, which has long been felt and acknowledged, of a scientific practical treatise on the blowpipe. Some explanation, however, and to the author some apology perhaps is necessary, as to certain liberties I have taken with the original. In the first place, I found the description of apparatus so very minute, that though such may be desirable in Sweden, in Britain I am sure it is not wanted, abounding as this country does in skilful artists, from whom every species of philosophical apparatus may be had, of the best workmanship and construction [1]. I have therefore

[1] Mr. Newman, of Lisle-street, Leicester-square, makes complete sets of apparatus for the blowpipe, containing every thing that the operator can require. Each article may be had separately, or the whole together, neatly arranged in a small case, with fluxes, &c. of the greatest purity, (an object of primary consequence,) at the option of the purchaser. Minerals of all kinds, both for collections and experiment, may be had of Mr. Sowerby, King-street, Covent Garden, and of Mr. Mawe, near Somerset-house, in the Strand.

shortened several of these descriptions, without however omitting any thing essential. In all that relates to the habits of the various subjects of experiment, whether of the substances in their pure unmixed state, or of their compounds as minerals, I have most cautiously abstained from deviating from my author; not a single character, in any one instance, has been intentionally omitted. But though I have held it matter of conscience to leave out nothing on these heads, I have been less scrupulous with respect to his nomenclature and formulæ. Berzelius, as is well known, has adopted an hypothesis of his own framing, founded on electro-chemical attractions, in which, when two elements combine, one of them is always supposed to be electro-positive, and one electro-negative, with respect to the other; and he has further laid it down as an established canon, that " compound atoms of the first order," (that is, composed of only two simple elementary atoms,) " having a common electro-negative element, always combine in such proportions, that the number of atoms of the electro-negative element of one, is a multiple by a whole number of that same number in the other; —that is to say, for instance, in the combinations of oxidated bodies, the number of atoms of oxygen of one of the oxides is a multiple by a whole number of that of the atoms of oxygen of the other; and in combinations of sulphurets, the number of atoms of sulphur of the one, is equally a multiple of the number of atoms of sulphur in the other."
(*Essai sur la Theorie des Proportions Chimiques,*

p. 37.) According to this canon, he calculates the composition of natural minerals, and expresses the results by certain signs and formulæ which he has invented for that purpose. With respect to his canon, it seems to me to differ in nothing essential from the common doctrine of chemical proportions, (in fact it is merely an hypothetical extension of it,) by which the greater number of compounds are illustrated with beautiful simplicity; and as to those which appear to be anomalous, it is probably better to let them stand for the present as acknowledged difficulties, than, in attempting to surmount them by arbitrary assumptions, incur the danger of involving the whole theory in obscurity, if not in error. Time, that has done so much, may do more; and future experiments will perhaps satisfactorily illustrate what now appears puzzling and obscure. When I turn to my scale of chemical equivalents, I find iron opposite the number 34·5, denoting its equivalent or ratio of combination, (that of oxygen being taken as 10,) oxide of iron 44·5, and red oxide of iron 49·5. This is a simple statement of facts, derived from the best authorities, and surely it is better to let it stand so (which answers every really useful purpose, and more perfectly than any other, because more simple), than to perplex ourselves with questions of atoms and half atoms, and by doubling this, and halving that, endeavour to make nature bend to our preconceived opinions. The author's nomenclature is closely connected with his canon; thus we find him expressing the composition of magnetic iron-ore (Traité, p. 145) as that of one atom of oxidule (protoxide) of iron, with two atoms of oxide (peroxide), the former containing two atoms of oxygen, the latter three; and, adopting Latin names and terminations, he calls the protoxide of iron, *oxidum ferrosum,* the peroxide *oxidum ferricum,* and the compound we are speaking of, *oxidum ferroso-ferricum.* This, and similar terms, I have taken the liberty to reject, using instead those commonly adopted in this country. The chemical and mineralogical symbols are composed of the initial letter or letters of the Latin names of the various elementary bodies; for the former, the common Roman capitals are used, over which are placed as many dots as there are supposed to be atoms of oxygen combined with the electro-positive element; for the mineralogical symbols Italic capitals are employed, and the dots omitted. In the formulæ they are connected by the usual algebraic sign of addition. These formulæ I have omitted in toto; but that the reader may not lose the information they are intended to convey, I have subjoined, in notes, the compounds they respectively indicate in common language. I have taken this liberty (and, I here beg to assure my author, with any feeling rather than that of disrespect), because I do not think the introduction of these, or any other symbols, at all necessary: it requires some time and patience to make oneself thoroughly master of

instance, he states the protoxides of columbium, copper, gold, mercury, and several others, to be formed of one atom of base and one atom of oxygen, whilst baryta, lime, silver, &c., having more energetic affinities, contain two atoms of oxygen to one of base. Berzelius, with Thomson, adopts oxygen for his unit, and in those instances in which he considers the lowest degrees of oxidation to require but one atom of that element, his numbers for the bases also correspond with Thomson's. Thus the atoms of columbium, copper, gold and mercury in Thomson's table are respectively 18, 8, 24·875, and 25; and in Berzelius's 18·23, 7·91, 24·36, and 25·31. But, with respect to the more energetic bases, Berzelius's numbers are double those of Thomson's; for, having laid down the before-mentioned hypothesis, there was no possibility of supporting it but by the *rule of two*. For instance he tells us (p. 149, Essai) that recent correct experiments show that 100 parts of magnesia contain 38·7 oxygen, and consequently are combined with 61·3 parts of magnesium: but magnesia, is one of the energetic bases, and *therefore* contains two atoms of oxygen; hence $\frac{38·7}{2}$: 61·3 :: 1·00 : 3·16, which is double Thomson's number, who with the generality of chemists (all of the present day, for aught I know, except Berzelius) considers magnesia as containing only one atom of oxygen. Berzelius has the same number to represent sulphur, as Thomson, (within a trifling fraction) and consequently is compelled to consider the sulphurets of the energetic bases as containing, in like manner, two atoms of sulphur, in order to reconcile experiment and hypothesis.

It is in these instances that I have committed the errors alluded to, and though it certainly is no excuse for my fault, yet I must take the liberty to say that it is one proof amongst a thousand of the danger of involving plain matter of fact in unnecessary hypothetical dogmas.

Having given this explanation I proceed to point out the corrections which the reader will have the goodness to make, by substituting the following numbers, for those he will find in the cases referred to. I am sorry that the lateness of the period, when the mistake was detected, should have made the list so numerous. The references are all to the notes.

P. 135. 1 atom of sulphur 16 + 1 atom of arsenic 38 = 54, and 3 atoms of sulphur 48 + 2 atoms of arsenic 76 = 124. The second compound is a sesquisulphuret, containing half as much more sulphur as the first; to express it *atomically* therefore the numbers are doubled to get rid of the anomalous half atom.

P. 136. 1 atom of arsenic 38 + 2 of oxygen 16 = 54.

P. 138. 1 atom of antimony 45 + 1 of sulphur 16 = 61, and the red, 1 atom of oxide of antimony 53 + 2 atoms of sulphuret 122 = 175. Thomson's and Berzelius's numbers for antimony do not accord well—that of the latter is nearly three times that of the former. The oxides and sulphurets of antimony are, like those of arsenic, not well made out.

P. 146. 1 atom of silver 110 + 1 atom of sulphur 16 = 126.

2 atoms of sulphuret of antimony 122 + 3 atoms of sulphuret of silver 378 = 500.

P. 149. 1 atom of silver 110 + 1 atom of mercury 200 = 310.

P. 150. 1 atom of bismuth 71 + 1 of sulphur 16 = 87.

P. 153. 1 atom of lead 104 + 1 of sulphur 16 = 120.

P. 155 1 atom of sulphuret of lead 120 + 1 atom of sulphuret of copper 80 + 1 of sulphuret of antimony 61 = 261.

P. 156. 1 atom of sulphuret of lead 120, 1 of sulphuret of silver 126, 1 of sulphuret of antimony 61, and 1 of arseniuret of nickel 64 = 371.

P. 161. 1 atom of oxide of lead 112 + 1 atom of tungstic acid 120 = 232.

P. 163. 2 atoms of sulphuret of copper 160 + 1 sulphuret of silver 126 = 286.

P. 165. 1 atom of sulphuret of tin 75 + 2 of sulphuret of copper 160 = 235.

P. 166. 1 atom of sulphuret of lead 120 + 2 of sulphuret of copper 160 + 2 of sulphuret of bismuth 174 = 454.

P. 167. 2 atoms of seleniuret of copper 210 + 1 of seleniuret of silver 151 = 361.

P. 173. $(\overline{80 \times 3} + \overline{16 \times 6}) = 336 + 12$ water $108 = 444$. Berzelius's number for silica is three times that of Thomson's.

P. 175. 1 atom of nickel $26 + 1$ of sulphur $16 = 42$.

P. 176. 1 atom of bisulphuret of nickel $58 + 1$ of biarseniuret of nickel $102 = 160$.

P. 177. For $\overline{26 \times 3} + \overline{62 \times 2} = 202$, read $\overline{34 \times 3} + \overline{62 \times 2} = 226 + 162 = 388$. The weight of the atom of nickel has been taken by accident for that of its oxide.

P. 183. 1 atom of uranium $125 + 1$ atom of oxygen $8 = 133$. The peroxide may be considered as containing half as much more oxygen as the protoxide, in which case it will consist of 2 atoms of uranium $250 + 3$ of oxgen $24 = 274$. Brande considers the atom of uranium to be 64, (oxygen being 8) and that the protoxide contains 1 atom of oxygen, the peroxide 2 atoms.

SKETCH

OF

BERZELIUS'S

MINERALOGICAL ARRANGEMENT,

BY THE TRANSLATOR.

As Berzelius's system of mineralogy is not generally known, I believe, in this country, I subjoin a sketch of its leading features. His arrangement of minerals is purely chemical; and although he may perhaps have proceeded somewhat more rapidly than is consistent with the actual state of chemistry, I am convinced that the basis is solid, and the superstructure will one day be perfect and beautiful. When I say that his system is *purely chemical*, it must not be understood that Berzelius rejects the use of external characters; on the contrary, he knows and acknowledges their value. It has been asserted that the *chemist* and the *mineralogist*, properly so called, *have*, and *ought to have*, different modes of viewing inorganic nature. This Berzelius expressly denies, and argues, that if the chemist, after relating the composition of a mineral, were to omit to describe, with the same care and accuracy, its external characters, as colour, hardness, transparency, crystalline form, &c. &c. his chemical details would be of little value, since no one, for want of the latter, could tell to what substance the former relates. But the principles on which a *general systematic arrangement* is to be founded, and those on which we are to proceed in the description of individual substances, are distinct from each other, and must not be confounded. As external characters are insufficient to determine the true place which a mineral, not as yet analysed, should occupy in the system; so, on the other hand, after an individual of any species has been analysed, they are essentially necessary to determine the identity of others, whose similarity, in these respects, to the one of ascertained composition, leaves no doubt of its being formed of the same elements, in the same proportions, and consequently obviates the necessity of a chemical examination of more than one well characterised specimen of each species. Hence, external characters are of great importance to mineralogy; but nevertheless, it is impossible to adopt a compound principle of general arrangement founded on the union of chemical composition and external form. The latter depends solely on the former, but great differences often occur in the external characters of minerals composed of the same elements, but in different proportions; and it not unfrequently happens that substances so circumstanced are even more dissimilar in point of external form, than others whose composition is less alike. Here composition and form are contradictory; and if the latter is to have part in the principle of general arrangement, we must transfer the mineral in such case from its association with its fellows, to those which it merely resembles externally. But no system attempts to proceed without founding its orders, at least on a chemical basis, employing the external characters in the subsequent details. If, therefore, we adopt the latter for arranging substances together, which, according to the fundamental principle, have no connexion, the whole system becomes inconsistent. Hence we are compelled to employ and keep separate, either composition, or external character, as the basis of general classification, without suffering the one to influence or disturb the other. The chemical arrangement does not make mineralogy the less a branch of natural history, since the objects it contemplates, though destitute of life and organization, and consequently incapable of being classed on principles suitable to organized bodies, are still a portion of the great work of nature. Nor is there any contradiction in considering mineralogy as a part of chemistry, because it is at the same time a branch of natural history; on the contrary, it is essentially a part of both, and the more assistance it derives from the former, the more perfect it will become as a portion of the latter. As to the objection, that our chemical researches have not yet attained the precision necessary to apply and confirm the new theory in its full extent, it is unfortunately true; but that is no argument against the propriety of adopting it, for the sooner we begin to treat a science according to accurate notions, the sooner our researches will be properly directed, and the sooner they will attain their object.

Having thus given an outline of the arguments by which Berzelius defends the superiority of a purely chemical system of mineralogy, I proceed to his classification, and the peculiar views on which he has founded it.

The influence of electricity on the theory of chemistry extends to mineralogy, although not hitherto applied to that science; for the elements of which minerals are composed, as well as those of all other bodies, unite with forces proportionate to the differences that exist in their mutual electrical relations. Hence, one or more electro-positive, and one or more electro-negative ingredients,[1] must be found in every compound body; thus, if it be formed of oxides, for every ingredient which we call a base, another must act as an acid, although the latter, in its insulated state, may not have the sour taste and other properties by which acids, usually so called, are distinguished; such are silica and the oxides of titanium, columbium, and many other metallic oxides, so that all the immense series of earthy minerals may be classed

---

[1] Electro-positive bodies are those which, when separated from a compound by the voltaic apparatus, terminate to the negative pole; electro-negative bodies terminate to the positive pole.

after the same principles as salts. An ingredient which acts as an acid in one case, may act as a base in another, according as it is electro-negative or electro-positive, with respect to the substance it combines with; and consequently, in a combination of two acids, the weaker may serve as a base to the stronger.

The doctrine of chemical proportions, which has so much exalted chemistry as a science, must also give the same mathematical precision, if we may be allowed the expression, to mineralogy, and there are a vast number of analyses of mineral substances, whose results perfectly accord with chemical proportions. It is principally from the examination of that class of minerals in which silica acts as an acid, and which may be called *silicates*, that most light is thrown on the other branches of mineralogy. Considered as an acid, silica forms several compounds of different degrees of saturation; the most common are the silicates, or those which contain one atom of silica, and one of base; sometimes two or three atoms of silica combine with one of base, forming respectively bisilicates and trisilicates; sometimes the base is in excess; for instance, two atoms of alumina may combine with one atom of silica, in which case the compound is a subsilicate. The order of arrangement depends on the electro-chemical properties of the elements of which mineral substances are composed, proceeding from the most electro-negative oxygen, to the most electro-positive potassium;

but as we are yet only very imperfectly acquainted with the electro-chemical relations of the simple bodies, we must be contented with an approximate arrangement.

Berzelius divides simple bodies into three classes: *Oxygen,—simple combustibles not metallic,—*(which he calls metalloids,) *and metals;* and they are distributed in each class, according to the order just mentioned; which is as follows:—

1. Oxygen.	Gold,
2. *Metalloids,*	Rhodium,
Sulphur,	Palladium,
Nitric,	Mercury,
Muriatic Radical,	Silver,
Boron,	Lead,
Carbon,	Tin.
Hydrogen.	Nickel,
3. *Metals.*	Copper,
Arsenic,	Uranium,
Chrome,	Zinc,
Molybdena,	Iron,
Tungsten,	Manganese,
Antimony,	Cerium,
Tellurium,	Yttrium,
Silicium,	Glucinum,
Columbium,	Aluminum,
Titanium	Magnesium,
Zirconium,	Calcium,
Osmium,	Strontium,
Bismuth,	Barium,
Iridium,	Sodium,
Platina,	Potassium.

" The *metals* are subdivided into electro-negative, or those whose oxides rather act as acids than as bases, and electro-positive, whose oxides act in preference as bases; and the latter are again separated into two subdivisions, the first of which contains those metals whose oxides are reduced by charcoal in the common manner, and the second those whose oxides cannot be so reduced. The latter are the bases of the earths and alkalies.

Each of these simple bodies forms a mineralogical family, composed of the simple body, and all its combinations with other bodies which are electro-negative with respect to it; that is, with those which, except in a few cases, stand above it in the preceding table.

The families are divided into orders, according to the different electro-negative bodies with which the electro-positive is combined. The orders are, for instance, 1. sulphurets; 2. carburets; 3. arseniurets; 4. tellurets; 5. oxides; 6. sulphates; 7. muriates; 8. carbonates; 9. arseniates; 10. silicates, &c. It is obvious that the number of orders increases as we approach the positive end of the series. The orders, when not very extensive, may be simply divided into species, which, according to Haüy's acceptation of the term, include those minerals that have the same composition, with the same primitive form; and the secondary forms which a species may present, constitute the *varieties*. But if the order be very extensive, and contain a great number of species, they are first separated into the following subdivisions: 1. salts composed of two ingredients, or simple salts; 2. salts composed of three ingredients, or double salts; 3. salts with three or four bases, triple or quadruple salts.

These subdivisions may in their turn be divided into genera, which comprehend all the minerals formed of the same ingredients, and these genera again into species.

In determining to what family a mineral belongs, somewhat different principles must be observed for the orders of *combustible*, and *oxidated* bodies.

For instance, if the mineral be a double or complex *sulphuret*, or *arseniuret*, &c., it must be placed in the family of the *electro-positive ingredient*, of which it contains the greatest number of molecules, or if the number of each kind be equal, in the family of the *most electro-positive*. If it be an oxidated mineral, composed of two or more oxides, it must always be placed under the most *electro-positive oxide*, without any regard to the number of molecules.

By attending to these two circumstances we obtain the great advantage of having minerals of analogous composition placed near each other, and the whole of the great class of double, triple, and quadruple silicates, arranged under the three or four last electro-positive substances, which terminate the system.

Each individual species is composed of the same ingredients united in the same proportions. Every variation in those proportions, or the smallest addition of any substance essentially belonging to the compound, produces a new species. Thus anhydrous gypsum forms one species; gypsum containing water of crystallization another: stilbite and chabasie are separate species, notwithstanding the slight difference in the proportions of their ingredients.

A mineral which contains a foreign mixture is placed under that species of which it has the most distinguishing characters, unless it assume the crystalline form of some other substance of which it contains but a few parts per cent. Thus under carbonate of lime are placed all the crystallized mixtures of that species, which contain carbonate of iron, and of manganese; but carbonate of iron, containing only five or six per cent. of carbonate of lime, is placed in the family of iron, although it seem to be moulded in the crystalline form of carbonate of lime. Were we in such cases to attend only to the crystalline form, it would divert us from the principle of the system, which is founded on the composition of the substance, whose true place we are looking for, and not on accidental circumstances, such as a foreign figure impressed on it by some unknown cause, &c.

I shall subjoin in this place two of the examples of the arrangement of substances in families given by Berzelius; the first in the family of silver, the second in that of aluminum. They are rather sketches to shew the method than perfect examples.

FAMILY: *Silver.*

First Order: *Native silver, in all its varieties.*

Second Order: *Sulphuret of silver.*
First Species: *Bisulphuret of silver.*
Second Species: *Sulphuret of silver, antimony, and iron.*
Third Species: *Sulphuret of silver and antimony, with oxide of antimony.*

Third Order: *Antimoniurets.*
First Species: *Sub-antimoniuret of silver* (antimonial silver).
Second Species: *Sub-tri-antimoniuret of silver.*

Fourth Order: *Tellurets.*
First Species: *Bitelluret of silver, with sex-telluret of gold* (graphic gold).
Second Species: *Bitelluret of silver, with bitelluret of lead, and tritelluret of gold* (weisserz).

Fifth Order: *Aururets.*
First Species: *Biaururet of silver* (electrum).
Second Species: *Subaururet of silver* (auriferous silver).

Sixth Order: *Hydrargyrurets.*
One Species: *Bihydrargyruret of silver* (native amalgam).

Seventh Order: *Carbonates.*

One Species: *Carbonate of silver.*

Eighth Order: *Chlorides.*

One Species: *Chloride of silver.*

The family of aluminum has no order belonging to the combustible compounds, nor perhaps any pure oxide, that is, alumina in an uncombined state. For even in the saphire, there are three and a half per cent. of silica, and seven per cent. of that substance in the ruby, though, in both, it possibly may be an accidental ingredient.

FAMILY: *Aluminum.*

First Order: *Sulphates.*

One Species: *Subsulphate of alumina* (native alumina from Halle and Newhaven).

Second Order: *Fluates.*

First Species: *Sub-fluate of alumina* (wavellite).

Second Species: *Fluate of alumina and soda* (cryolite).

Third Order: *Fluo-silicates.*

One Species: *Fluo-silicate of alumina* (topaz).

Fourth Order: *Silicates.*

First Sub-division: *Simple silicates.*

First Species: *Silicate of alumina* (nepheline).

Second Species: *Tri-sub-silicate of alumina* (collyrite).

Second Sub-division: *Double silicates.*

First Genus: *Silicates, with base of glucina and alumina.*

First Species: *Bisilicate of alumina, with quadrisilicate of glucina* (emerald).

Second Species: *Its exact composition not determined* (euclase).

Second Genus: *Silicates, with base of lime and alumina.*

First Species: *Trisilicate of lime, with bisilicate of alumina* (mealy stilbite).

Second Species: *Bisilicate of alumina and lime* (laumonite).

Third Species: *Silicate of alumina and lime* (vitreous paranthine).

Third Genus: *Silicate, with base of alumina and baryta.*

One Species: *Quadrisilicate of baryta, with bisilicate of alumina* (harmotome).

Fourth Genus: *Silicates, with base of soda and alumina.*

First Species: *Trisilicate of soda, with silicate of alumina* (mesotype).

Second Species: *Silicate of soda and alumina* (tourmaline apyre).

Fifth Genus: *Silicates, with base of potassa and alumina.*

First Species: *Trisilicate of alumina and potassa* (feldspar).
Second Species: *Trisilicate of potassa, with bisilicate of alumina* (meionite).
Third Species: *Bisilicate of alumina and potassa* (amphigene).

Third Sub-division: *Silicates with triple, and quadruple bases.*

First Genus: *Garnets with a triple base.*

First Species: *Bisilicate of protoxide of iron, silicate of protoxide of manganese, and silicate of alumina* (garnet from Broddbo).
Second Species: *Same composition* (garnet from Finbo).
Third Species: *Silicate of protoxide of iron and manganese, and silicate of lime and alumina.*

Second Genus: *Different kinds of mica.*

First Species: *Trisilicate of potassa, with silicate of peroxide of iron, and silicate of alumina* (foliated mica).
Second Species: *Similar composition, but in different proportions* (silvery mica).
Third Species: *Trisilicate of potassa, with si-*

licate of magnesia, and alumina, and iron (black mica).

Berzelius calls those minerals, which, like most of the rocks, consist of combinations, evidently dissimilar on mere inspection, *mixed;* when the eye cannot distinguish the dissimilar parts, and they are not disclosed either by the fracture, or by cutting and polishing the surface, he considers them as having been *fused together,* but not as chemically united. *Pure or unmixed minerals* are those whose constituent parts are united in such proportions as to form true chemical compounds, and *simple minerals* contain only one element; diamond is a simple mineral, emerald a pure unmixed mineral.

It is the business of chemical analysis to discriminate between mixed minerals, and those which are chemical compounds, and this the atomic theory enables it satisfactorily to accomplish, at least in most cases. It does not follow, that because a mineral is crystallized, it must be absolutely pure. Crystals often contain some portion of foreign matter, derived from the liquid from which they have separated, or from other causes; in these cases, therefore, not less than in those that are amorphous, we must apply the same theory to distinguish between the real chemical compounds, and the foreign substances accidentally present. To this end, the *forces* by which those compounds are formed, must be kept in view. As these forces depend on the electrical polarities of the combining particles, they can only unite in pairs, since there cannot be a third co-operating power. Sodium, an electro-positive body, combines with oxygen, an electro-negative body, and together they form a compound, in which their original polarities disappear. Sulphur in the same manner combines with oxygen, and forms sulphuric acid. But though the original polarities be neutralized, those two bodies, which must now be considered, respectively, as to their further combinations, as sole and elementary, are still electro-positive and electro-negative, with regard to each other; hence they also are capable of uniting chemically, and the compound they form is not a triple union of sodium, sulphur, and oxygen, but a binary one of soda and sulphuric acid. Again, this new body is capable of combining with water; and the crystallized salt must, in like manner, be considered as a binary compound of sulphate of soda and water. Suppose it were required to make a perfect analysis of crystallized alum, we must first separate the two ingredients of which it is immediately composed, anhydrous alum and water. These must next be separated, the latter into its elements, oxygen and hydrogen, the former into its immediate ingredients, sulphate of potassa, and sulphate of alumina, which again must be divided, each into its constituent parts, and so on to its ultimate elements; and all that applies to this example, applies equally to every other body composed on the principles of inorganic nature, namely, binary combination. Hence, when we examine the results of the analysis of a mineral, and compare the weight of each of its ingredients with the weights of their respective atoms, we discover, if there be any of them, whether electro-positive or negative, for which a corresponding one of the other order, either in an equal, or in some multiple, or submultiple proportion be wanting; in which case, such ingredient must be considered as accidental, and not as forming an essential part in the composition of the mineral. For instance, Hisinger's analysis of malacolite, gave him,

        Silica................ 54·18
        Lime................ 22·72
        Magnesia .... 17·81
        Oxide of iron......... 2·18
        Oxide of manganese.. 1·45
        Volatile matter ...... 1·20
                                  ——
                                 99·54

Here, three of the ingredients enter in large quantity into the composition of the mineral; the others in very small quantity. It is natural to suppose, therefore, that the former are the true elements of the compound, and since silica acts as an acid in such combinations, that malacolite may be a true silicate of the bases, lime, and magnesia. To ascertain this, we must examine whether these ingredients be in such proportions as are necessary to form those salts or not. The weight of an atom of

$$\text{Silica is} \ldots\ldots\ldots\ldots 16$$
$$\text{Lime} \ldots\ldots\ldots\ldots\ldots 28$$
$$\text{Magnesia} \ldots\ldots\ldots\ldots 20$$

Now 28 : 16 :: 22·72 : 13
And 20 : 16 :: 17·81 : 14·24

The quantity of silica which the lime and magnesia require to form silicates, we thus find is 27·24; but that given by analysis is 54·18. Now this number is almost exactly double the former, or within a trifling fraction just that quantity which the two bases require to form bisilicates. This coincidence therefore justifies the conclusion, that malacolite is a bisilicate of lime and magnesia, and that the iron and manganese are not essential to its composition, nor chemically united to the other ingredients. Having thus ascertained what are the true chemical elements of the mineral, we next inquire, how many of the atoms of each it may probably contain. The solution of this question is very simple. Suppose the proportions had been by the analysis (disregarding the non-essential ingredients)

$$\text{Silica} \ldots\ldots\ldots\ldots 55\cdot43$$
$$\text{Lime} \ldots\ldots\ldots\ldots 16\cdot17$$
$$\text{Magnesia} \ldots\ldots\ldots 23\cdot10$$

and by the preceding method we had found the

composition of the mineral to be as before a bisilicate of lime and magnesia, and the proportion of the former to the latter, as 34·67 : 60. The weight of an atom of bisilicate of lime is (28 + 32) 60, and that of an atom of bisilicate of magnesia (20 + 32) 52. Therefore as 60 : 52 :: 34·67 : 30 = $\frac{1}{2}$ of the quantity of bisilicate of magnesia. Consequently the most simple estimate of the composition of the mineral will consider it as containing 1 atom of bisilicate of lime + 2 atoms of bisilicate of magnesia. In the same manner the atomic constitution of all other minerals, however complex, may be ascertained.

INTRODUCTION

BY

THE AUTHOR.

The subject of the work now offered to the public, is highly interesting to the practical chemist, the miner, and the mineralogist. It is a system of chemical experiments, made in the dry way, as it used to be called, and almost always on a microscopic scale, but which presents us in an instant with a decisive result.

In the analysis of inorganic substances, the use of the blowpipe is indispensable. By means of this instrument, we can subject portions of matter, too small to be weighed, to all the trials necessary to demonstrate their nature; and it frequently even detects the presence of substances not sought for nor expected in the body under examination. The facility that it affords for discovering the constituent parts of metallic fossils, renders it equally indispensable to the miner, whose common processes are sometimes singularly disturbed by the occurrence of foreign substances in the minerals he operates on, and whose nature, for want of time or skill, he can but seldom ascertain by sufficiently elaborate and delicate chemical experiments, but which the ready and convenient use of the blowpipe enables him to develope in a few seconds. To the mineralogist this instrument is absolutely necessary, as his only resource for immediately ascertaining if the inferences he draws from external characters, such as form, colour, hardness, &c. be legitimate. In the following pages the series of results, to which researches of this kind are capable of leading us, are detailed. I have successively described the phenomena which the various subjects of the mineral kingdom present when submitted to the test of the blowpipe, by experiments made, as far as possible, on pure and well characterised specimens. I have carefully pointed out the localities of the several minerals, whenever I suspected that differences in their beds might lead to discrepancies in the results, in order that those discrepancies may be distinguished from such as might arise from inaccurate observations recorded in this work. As to errors of this kind, I could not wholly avoid them, without multiplying my observations; and the number of minerals which I had to examine was too great, to allow me to repeat in sufficient detail, all the experiments relating to each individual.

I have quoted the names of those persons to whom I am indebted for the rarer substances, desirous of thus giving at once a guarantee of the accuracy of the mineralogical names I have adopted, and of acknowledging the liberality with which they have supplied me with specimens of the rarest minerals for my researches. I seize this opportunity to mention with gratitude the names of Haüy, Bournon, Gillet de Laumont, Brongniart, Brochant, Cordier, Lucas, Weiss, and Blöde.

When we meet with a mineral not sufficiently distinguished by its physical characters, there are generally but few known fossils with which it can be confounded. If, therefore, we compare the effects produced by the action of the blowpipe on the mineral in question, with the descriptions of analogous effects on the minerals which resemble it, we shall seldom fail to ascertain the nature of the one we are operating on. As to the metallic fossils, nothing is easier. With the purely earthy minerals the result is not quite so decisive; since, in experiments with the blowpipe, the precision of the result depends in general rather on differences in the *nature* of the constituent parts of a mineral than on their *proportions*. But even in compounds which differ only in the relative quantities of their component parts, we often find such remarkable differences in their degrees of fusibility, or the phenomena which they present when treated by the same flux, that the blowpipe is still sufficient to decide the question.

As we are often mistaken respecting the nature of a mineral, when, considering only its external character, we disregard the phenomena which depend on its chemical composition, so also, we seldom succeed in determining it, if, neglecting the former, we rely wholly on the effects produced by the blowpipe. Those modern mineralogists who think that an inorganic substance may always be known by its appearance, are compelled, in many cases, to have recourse to the blowpipe, (supposing them to have sufficient chemical knowledge to use it with advantage,) when they find themselves led into error by principles founded on vague perceptions and imperfect data.

A mineralogical classification has been attempted founded on the effects of the blowpipe, as a means of ascertaining the genus and name of any mineral whatever, on principles analogous to those which regulate the classifications of organic nature. Aikin's Mineralogy is a very ingenious attempt of this kind, and as a proof of the sagacity of its author, I think it right to mention, on this occasion, that he has not confounded the kind of classification in question, with the systematic arrangement on which the science, properly so called, proceeds. Now this the mineralogical school, which still prevails in Germany, does, in attempting to mould the second on the first. However, I do not believe, that a classification of minerals, founded on the phenomena they present before the blowpipe, sufficiently accomplishes its object to enable a person well skilled in the use of that instrument, but ignorant of mineralogy, to ascertain a mineral submitted to his examination.

I have adopted the chemical system for the classification of minerals in the following work; and in certain places, as in the case of amphibole, pyroxene, and garnet, I have indulged in some theoretical remarks, founded on the consideration that pyrognostic operations are the more interesting, in proportion as we are better acquainted with the nature of the bodies submitted to their test.

Lastly, I have briefly shown how to determine, by the help of the blowpipe, the constituent parts of the stony concretions which form in the urinary passages. Being often consulted by medical men, as to their composition, I have been obliged to seek the shortest mode of obtaining a result. I now publish what I have learnt from experiment on the subject, in order to put the medical practitioner in the way of making such analyses without the assistance of the chemist.

HISTORY

OF

THE BLOWPIPE.

JEWELLERS and other workers in metal on a small scale, avail themselves of the use of the blowpipe, to direct the flame of a lamp on pieces of metal supported on charcoal, so as to fuse the solder by which they are to be united.

This instrument was long employed in the arts, before any one conceived the idea of applying it to chemical experiments, performed in the dry way, as it is called.

Bergman tells us, that the first person who so used it was Andrew Swab, a Swedish metallurgist, and Counsellor of the College of Mines, about the year 1733. He left no work on the subject, and it is unknown to what extent he carried the researches he made with this instrument. Cronstedt, who laid the foundations of mineralogy, and whose genius so outstripped the age in which he lived, that he was unintelligible to his cotemporaries, used the blowpipe to distinguish mineral substances from one another, by the means of fusible reagents, whose action should produce such modifications on the objects to which they were applied, as might afford some conclusions respecting their composition, and serve as a basis for the classification he adopted. In his time the intercourse between men of science was by no means so open as at present; the labours of one man were seldom communicated to his fellow labourers, and each pursued his researches with no other help than the experience of the generation which had passed away, and become in some measure public property. At such a period, Cronstedt carried the use of the blowpipe to a degree of perfection that can only result from persevering industry; but as slight services in the cause of science were not as yet honoured with general attention, he who at first was afraid to make himself known as the author of that system of mineralogy which has perpetuated his memory, still less thought of describing in detail this new application of the blowpipe, and the processes he adopted. He only published such results of his experiments, as might serve to distinguish minerals from each other, by affording characters peculiar to the different species.[1]

Von Engeström, who published an English translation of Cronstedt's system in 1765, annexed to it a treatise on the blowpipe, in which he par-

---

[1] The first edition of his work appeared in 1758. B.

ticularly noticed his (Cronstedt's) processes, as well as the principal results of their application to the minerals then known. This treatise did not appear till 1770,[1] and was translated and published in Swedish, by Retzius, in 1773. This work attracted the general attention of chemists and mineralogists to the use of the instrument, who, however, derived at first little other advantage from it, than as a means of ascertaining the fusibility of bodies, and occasionally their solubility in glass of borax; for the want of that skill in its application, which can only be derived from patience and practice, together with a sufficient knowledge of the phenomena presented by the various substances used as fluxes for the bodies experimented on, prevented a just estimate being formed of its value, whilst the difficulties attending its use were abundantly evident; and hence every where but in Sweden, the art of the blowpipe made but little progress. As in other practical sciences, books alone are "weak masters" to make adepts in this; but they who had seen Cronstedt and Von Engeström at work, learned to work like them, and transmitted their skill to their successors. Bergman went further than Cronstedt; he extended the use of the blowpipe beyond the bounds of mineralogy, to the field

---

[1] An Essay towards a System of Mineralogy, by Cronstedt, translated from the Swedish by G. Von Engeström, revised and corrected by Mendes da Costa, London, 1770. B. The best edition is that by John Hyacinth de Magellan, 2 vols. 8vo. Dilly. 1788. C.

of inorganic chemistry; in his hands it became an invaluable instrument for discovering very minute portions of metallic matter, in analytical researches. His work De Tubo Ferruminatorio was first published at Vienna, under the direction of D. A. Born, in 1779.[1]

Bergman, on account of his health, was assisted in his experiments by Gahn, who particularly applied himself to the use of the blowpipe in his mineralogical studies, in consequence of the readiness with which it affords decisive results. The operations which he performed under Bergman's inspection, who caused him to examine all the minerals then known, taught him how each individual conducts itself before the blowpipe. Assisted by the experience he thus acquired, he continued to employ the instrument in every kind of chemical and mineralogical enquiry, whence he attained to such a degree of skill in its use, that he could detect the presence of substances in a body by its means, which had escaped the most careful analysis, conducted in the moist way. Thus, when Ekeberg asked his opinion respecting the oxide of *Columbium*,[2] then recently discovered, and

---

[1] The English reader will find it in the second volume of Bergman's Physical and Chemical Essays, translated by Dr. Cullen, and published in London, 1788. C.

[2] *Oxide de Tantale.* I have restored the original name given to this substance by its discoverer, Mr. Hatchett, and the only one by which, in England at least, it ought to be distinguished. Mr. Phillips, in his excellent " Introduction

of which he sent him a small specimen, Gahn immediately found that it contained tin, although that metal does not exceed 1-100th of the weight of the mineral. Long before the question was started whether the ashes of vegetables contain copper, I have seen him many times extract with the blowpipe, from a quarter of a sheet of burnt paper, distinct particles of metallic copper.

Gahn always travelled with his blowpipe, and the continual use which he made of it, led him to several improvements in its application; he exa-

---

to Mineralogy," also ascribes the honour where it is due, using the terms *Columbium*, *Columbite*, and *Yttro-Columbite* as the generic, and those of Tantalite and Yttro-tantalite as their synonyms. Dr. Thomson, in the first volume of his Chemistry, sixth edition, under the class of simple combustibles, heads his fifth section with " Of Columbium, or Tantalum," and, having shewn Mr. Hatchett's right to name the result of his own labours, immediately adopts the appellation bestowed on it by its subsequent discoverer, M. Ekeberg, the identity of whose *Tantalum* with Hatchett's *Columbium* was fully demonstrated by Dr. Wollaston in 1809. (Phil. Trans.) In his description of minerals, in the third volume, Dr. Thomson adopts *Tantalite* as the leading name, and gives "Columbite of Hatchett" as its synonym. The confusion of terms in the first and second volume of that work, is diverting; e. g. " The mean of four experiments made by Berzelius, in which he oxidized determinate weights of *Tantalum* by means of nitre, gives the composition of *Columbic* acid as follows: *Tantalum*, 100," &c. vol. 1, p. 576. But we must add in justice, that Thomson's Chemistry is a work, in spite of many carelessnesses, of high value. C.

mined a great number of reagents, in order to find new methods of arriving at the knowledge of certain substances, and the whole was imagined and executed with such sagacity and precision, that his results were entitled to the greatest confidence. He most readily and carefully instructed those who were desirous of information on the subject, but he never appears to have thought of publishing an account of his labours, nor has it been done by others.[1]

I was so fortunate as to enjoy a familiar intercourse with this eminent man during the last ten years of his life. He spared no pains to impart to me all that he could from his knowledge and long experience, and I have strongly felt the obligation I then contracted towards the public, to perpetuate, as far as in me lies, the fruits of his labours. At my earnest request, he composed the most important parts of what I have published on the blowpipe, and its use, in the second part of my Ele-

---

[1] Mr. Professor Hausmann is the only person who has given an account, at any length, of the services which Gahn's blowpipe has rendered to science, in a memoir inserted in the *Mineral-Taschenbuch de Leonhard*, for the year 1810. B. There is in the eleventh volume of the Annals of Philosophy, p. 40, an article "On the Blowpipe; from a Treatise on the Blowpipe, by Assessor Gahn, of Fahlun." The greater part of that treatise, together with some additional matter on the subject, may also be found in the appendix to my translation of the fourth volume of Thenard's Chemistry, entitled an "Essay on Chemical Analysis," p. 374. C.

ments of Chemistry;[1] it is all that we have from him on the subject. He never wrote any memoirs respecting the results of his experiments on minerals, and when, through age, his recollection of the facts he had observed began to fail him, he often insisted on the necessity of examining and carefully noting down the phenomena presented by different minerals when acted on by the blowpipe. At his desire, I undertook such an investigation, the results of which he proposed to criticise, his blowpipe in his hand, that where any discrepancies might exist, we should be enabled to detect their causes, and avoid all inaccuracy in the place assigned to each individual substance. This scheme was frustrated by his unexpected death, an event that happened too soon, long as his life had been.

In all the rest of Europe only one naturalist, but he a most distinguished one, has applied himself to the study of the blowpipe and its uses,[2] and submitted a large number of mineral substances to

---

[1] Larbok i Kemien, 2; dra Delen. Stockholm, 1812. p. 473, et seq. B. It is to be regretted that we have no English translation of this work. Its origin forbids a doubt of its value. C.

[2] I cannot subscribe to this opinion; for in this country alone it is well known there are many persons who have made great proficiency in the use of the blowpipe. *One* in particular, whom it is unnecessary to mention by name, is as eminently distinguished for his dexterity in managing this useful little instrument, as he is for the general accuracy of his conclusions, and the sagacity by which he arrives at

its test. This was H. de Saussure, justly celebrated for his geognostic researches on the Alps of Switzerland. He, as well as Cronstedt, employed it chiefly in distinguishing minerals, and although he made additions and improvements on the subject, he ranks far behind Gahn in respect to the results which he obtained with this instrument.

DESCRIPTION OF THE BLOWPIPE.

The blowpipe commonly employed in the arts, consists of a conical brass tube, about 13 inches long, 3-8ths of an inch in diameter at the larger end, and having an opening at the smaller of rather less than 1-16th of an inch. It is bent at a right angle, at about an inch from the point, and in using it the large end is placed in the mouth, and the blast directed on the flame of a lamp by the small end. (Pl. I. fig. 1). In the arts it is seldom necessary to continue the blast more than a minute, so that no inconvenience results from the vapour of the breath being condensed in the interior of the tube; but in chemical experiments, which often last a considerable time, this would become a

them. Had he thought it worth while to communicate his knowledge on the subject to the world, the present work would have been as unnecessary as the assertion is erroneous. C.

Fig. 1.

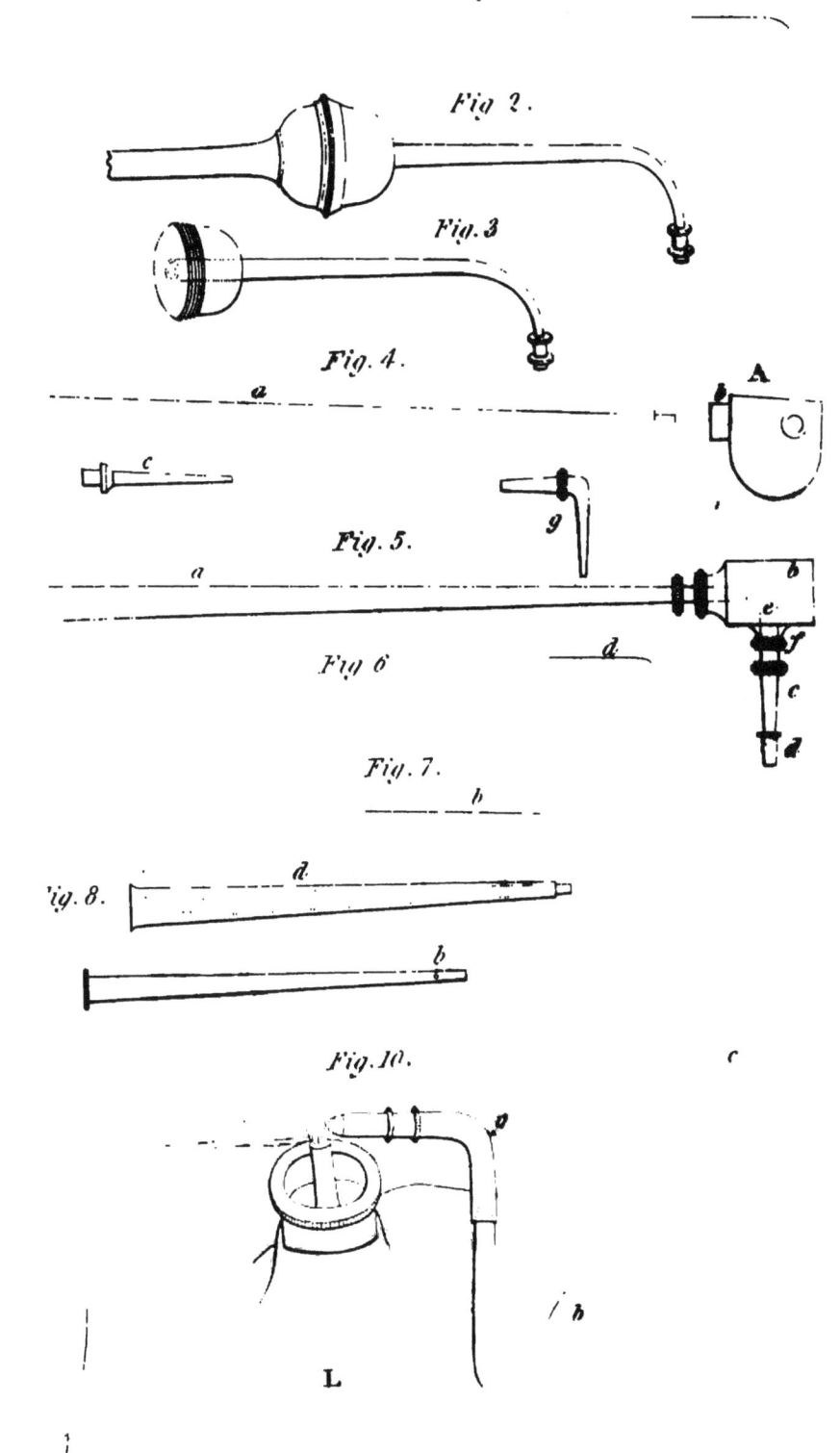

Fig. 2.

Fig. 3.

Fig. 4.

Fig. 5.

Fig. 6.

Fig. 7.

Fig. 8.

Fig. 10.

source of great annoyance. To obviate it, Cronstedt placed toward the small end of the instrument, a hollow ball, intended to collect the condensed vapour. This was a material improvement, but still had the inconvenience, that when held in a vertical position, the small end downwards, the water in the ball flowed into the beak, or small end of the instrument, and impeded the blast, or was projected into the flame.[1] To remedy this, Bergman adopted a semicircular chamber of the size and form seen fig. 4. A, as the receiver to his blowpipe, inserting the long tube, *a*, into the neck, *b* ; and the beak, *c*, into the hole at the upper part of one side of the chamber at *d*. It is obvious that the water collecting in the lower part of the receiver cannot find its way into the beak, *c*. Gahn gave his receiver the form of a cylinder, one inch long, and half an inch in diameter, as seen at fig. 5. In other respects, his instrument resembles Bergman's, and consists of four pieces, *a*, *b*, *c*, *d*. The little jet, *d*, is fitted by grinding to the

---

[1] I believe it was Mr. Pepys who found a remedy for this grievance, equally simple and effectual, by producing the tube of the beak so as to project on the inside of the ball, through about half its diameter, thus preventing any condensed water from passing out in that direction, at least till the ball is half full of it ; and this being made of two hemispheres united by a screw, it is taken to pieces and cleaned with little expense of time or trouble. Fig. 2. shows the appearance of this instrument when put together ; fig. 3. the beak with its hemisphere ; the dotted line shews the tube projecting on the inside. C.

extremity of the beak, *c*, and there should be two or three of these jets with holes of different sizes, to be changed as occasion requires. The advantage of Ghan's blowpipe over Bergman's is owing to the cylindrical shape of the receiver, which occupies less space than the former, and to the length and conical form of the neck, *e*, of the beak, inserted in the cylinder at *f*, which allows it to be pushed farther into the receiver, when worn by long use, and at all times causes it to fit tight, and not be liable to fall out of its socket during an experiment,—an inconvenience not unfrequent with the blowpipe contrived by Bergman. In my estimation Gahn's blowpipe is preferable to all others. Instead of the straight beak, *c*, I employ a bent one, *g*, when I have occasion to use the instrument for blowing glass; its rectangular elbow enables me to give it every possible direction with respect to the tube, *a*. As it is highly desirable, especially for the mineralogist, that the blowpipe should occupy the least possible space, and be very portable, without, at the same time, losing any essential quality, several chemists have endeavoured to obtain the limits of simplicity in its construction, —an object most effectually accomplished in the blowpipes of Tennant and Wollaston.

Tennant's blowpipe consists of a straight and very slightly conical tube, (pl. I. fig. 6. *a*, *b*.) closed at one end, at half an inch from which is an opening to receive the small bent tube, *d*, which is fitted in by grinding, and may be turned in any direction required. When used, the beak of the small tube generally makes a right angle with the tube, *a, b ;* when in its case, it lies parallel to it as in the figure. This instrument unites in a high degree every advantage that can be wished for with great simplicity; the water condensed from the breath flows to the closed end of the principal tube without entering the beak. Wollaston's blowpipe is still more portable than Tennant's. It is composed of three pieces of the size and form represented in *a, b, c.* (pl. I. fig. 7.) and when the instrument is not in use, the third piece is sheathed in the second, and the second in the first, so as to reduce its length to one half, and make it occupy no more space than a common pencil case. (See fig. 8.) The small end of *a* is fitted by grinding into the large end of *b*, whose opposite extremity is closed, but near it is a small lateral opening, as seen at *b*, fig. 9. The beak, *c,* is closed at the large end, and has a very fine hole for the blast at the point; it receives the piece, *b,* by a transverse opening, not perpendicular to its axis, but so inclined as to cause it, when the instrument is put together, to form with the main tube the convenient obtuse angle represented in fig. 7. For portability, this blowpipe far exceeds all others; it packs with perfect convenience in one's pocket book, and if we add to it a slip of platina foil, and a small piece of borax, we are furnished at once with a sufficient laboratory for a great variety of operations,—for the candle and charcoal may be found every where.

The length of a blowpipe, whatever be its form, must depend on the eye of the operator, and be such, that the body operated on may be at that distance from his eye at which he has the most distinct vision.

Blowpipes are best made of silver, or tinned iron plate, the beak only being of brass. If the instrument be wholly of brass, it in time acquires the odour and taste of verdigris,—an inconvenience not entirely removed by making the mouth piece of ivory. The hands, too, if not quite dry, contract the same odour during the operation, especially if the blowpipe has not been used for some time, and was not well cleaned before it was laid by. Tin plate is not liable to this nuisance, and has, besides, the advantage of cheapness. When that material is employed, the joinings should be made air tight by inserting pieces of paper between them. Notwithstanding that silver is the best conductor of heat of all the metals, no inconvenience need be apprehended on that score, even in the longest operations. The small jets adapted to the extremity of the beak are a great improvement, for the extremity is liable soon to become covered with soot, and the hole to be either blocked up, or lose its circular form, and it is necessary to clean it, and clear the opening with a small needle kept at hand for the purpose. This is an indispensible but troublesome operation. I have, therefore, had these jets made of platina, each of a single piece, and when dirty I heat them red hot on a piece of charcoal, which cleans them in an instant, and clears the hole without the assistance of any mechanical agent. Silver would not answer the purpose, for although we are ever so careful not to fuse it in the operation, it would crystallize as it cools after having been heated to redness, and become as brittle as an unmalleable metal.

Glass blowpipes are certainly less costly and less liable to get dirty than those made of metal, but their brittleness and the fusibility of their beaks are so serious inconveniences, that they should never be used but in cases of necessity.[1]

Attempts have been made at different times to construct blowpipes of greater facility in their management than the one in common use. De Saussure fixed his to a table, so as to have his hands at liberty, whilst he regulated the blast by his mouth. I know no case in which much advantage is derived from this plan; moreover, the variations in the sort of flame required in experiments with the blowpipe, depend on such slight changes in the position of the beak, that it is impossible to accomplish them with precision by the mere action of the mouth.[2]

---

[1] In this country the blowpipes are generally made of brass, and, when well lackered, I have never been annoyed with the inconveniences Berzelius complains of. The mouth-piece should be of silver, and the beak of platina; were it not for the expence, the whole instrument were better made of platina. C.

[2] These expedients are like the various devices for lathes

It is generally supposed, that to use the blowpipe is a difficult operation, that it requires great pulmonary exertion, and may be injurious to the health. Partly for this reason, and partly for want of skill to keep up a continued blast without inconvenience, various contrivances have been hit upon to supersede the blowing with the mouth. Such

and tools for *gentlemen* turners and carpenters, who waste their time and cut their fingers in ineffectual attempts to make a box worth sixpence, with an apparatus that cost a hundred pounds. The skilful workman needs no such aids, and the operator with the blowpipe will do well to render himself independent of them at once. However, as some readers may be of a different opinion, and as in a few cases where an unusually large flame is required the instrument may be useful, I annex a figure of the best form that I have met with, for a blowpipe on that construction. I do not know who is the author of the invention. a, (fig. 10, pl. I.) is a rectangular copper box, $2\frac{1}{2}$ inches long, 1 inch high, and $\frac{1}{4}$ of an inch wide, which is fixed on a board by a screw passing through the foot-plate, b; c is a tube projecting from the top of the box, to which one end of a flexible tube is adapted by a brass socket. The flexible tube is terminated at its other end by an ivory mouth-piece, and is of such a length, as to be conveniently applied to the mouth of the operator; d is another projecting tube, to which the beak, f, g, is adapted; and e is a third projecting tube (closed with a cork) for the purpose of pouring off the condensed water that collects in the box, a. The beak is made in two parts, f, and g, (fig. 11,) the large end of f fitting on the projecting tube, d, and g, in like manner, fitting on the small end of f. By this means a double rectangular motion is obtained, which allows the beak to be presented in any position to the flame of the lamp, l. All the parts that have motion are well fitted together by grinding. C.

are those of Ehrmann, Köhler, Meusnier, Achard, Marcet, Brooke, and Newman, by means of which the highest degrees of heat are produced, on a small scale, by an artificial blast of atmospheric air, or oxygen gas; but as the use of these instruments has no connection with our subject, I shall abstain from further notice of them.[1]

[1] I shall not: for in a work especially devoted to the blowpipe, some account of every important instrument should, in my estimation, be included; and although, in some respects, Brooke's or Newman's blowpipe (for it is but one instrument) is not calculated for mineralogical experiments, in others it is highly useful. Indeed, when used with atmospheric air, it *may* be applied to *all* mineralogical purposes, though still with less advantage than the common blowpipe in *skilful* hands. But when filled with a condensed mixture of oxygen and hydrogen gases, in the proportions requisite to form water, one essential character, the fusibility, or infusibility of different substances, as determined by the common blowpipe, disappears before the intense heat produced by this, which levels all bodies to one general class of fusible substances; though very evident differences are still observable in the *facility* with which different bodies are reduced to the state of fusion. In return, too, for the character which is thus lost, we gain a new one in the appearance of the, otherwise infusible, body, after it has been melted. For these reasons, a description of the instrument seems desirable, and a figure of it may be considered as a good substitute for the plan of a red morocco case, to hold mineralogical instruments, and the piece of furniture, which occupy the fourth plate in the original volume. This apparatus was first made at the desire of Mr. Brooke, by Mr. Newman, of Lisle-street. An accident that occurred to Dr. Clarke, by the explosion of the reservoir, and which had nearly been attended with serious consequences, occasioned several attempts at its im-

16   DESCRIPTION OF

Hassenfratz supplied his blowpipe with air by means of a pair of bellows worked by the feet,

provement, especially with regard to safety; the most perfect of which is the trough represented by fig. 4, pl. II. It was suggested by Mr. Professor Cumming, of Cambridge. *a*, (fig. 1, pl. II.) is the reservoir made of sheet copper, $5\frac{1}{2}$ inches long, 3 inches wide, and 3 inches high; *b*, a syringe connected by a couple of stop-cocks, *c*, to the reservoir; *d*, is the head of the trough (or safety apparatus), fitting in its place by a screw, perfectly air tight: the trough is inserted in the reservoir, in the direction of the dotted lines, and descends to the bottom: it is represented, on a large scale, at fig. 4.; *e*, a stop-cock proceeding from the head *d*, and *f*, its jet fixed to it by the ball and socket joint *g*. When the instrument is used, its parts are to be put together, as in fig. 1. and the reservoir exhausted, by working the piston of the syringe, *b*. The stop-cocks must then be closed, the syringe with the upper stop cock taken off, and the syringe alone placed in the upright position shewn at fig. 2. The bladder, *h*, containing the gases, must then be connected by the screw socket, *k*, and its stop-cock, with the syringe. The syringe stop-cocks are now to be opened, when the gases will issue from the bladder and fill the reservoir. The head of the trough is then to be unscrewed by the key, (fig. 3.) and oil poured in, to about half an inch above the lower screen of wire gauze, (see fig. 4.) and the head again screwed tight in its place. The gases are next to be condensed into the reservoir, by working the piston of the syringe as before, and all the stop-cocks being now shut, the apparatus is ready for use.

During the whole time the jet is burning the oil will be heard to play in the trough. If the current be inflamed, and the instrument abandoned to itself, the jet will go on burning until the expansive force of the atmosphere within the box is no longer sufficient to propel a stream with the required rapidity through the tube; at this time the inflamma-

Fig. 1.

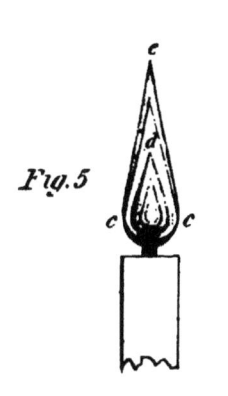

Fig. 5.

Fig. 6.

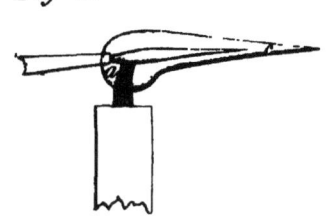

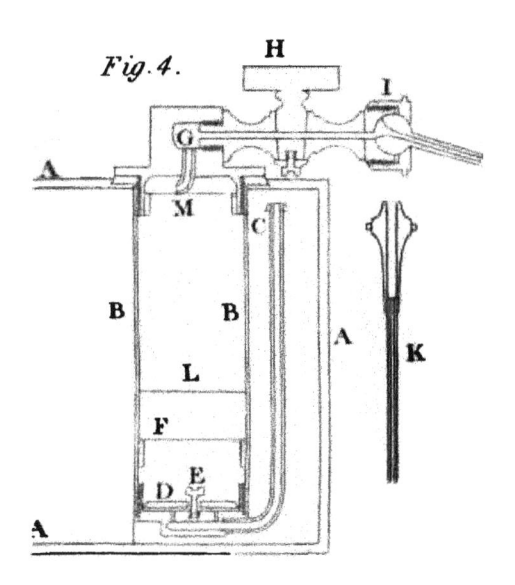

Fig. 3.

Fig. 4.

Fig. 2.

after the manner of the enamellers, and Näsen employed for the same purpose a bladder filled with

tion will pass backwards, unless the tube be very fine, and will fire the small quantity of mixture in the upper part of the trough, and then its effects will cease, the atmosphere in the reservoir remaining as before. When, however, the regular use of the instrument is required, it is better to shut the jet-cock before the atmosphere is quite out, and condense in a fresh portion of the gas.

Attention should be paid to the quantity of oil in the trough —it should cover the gauze, but not to too great a height; if there be too much oil, it is possible that the agitation caused by the passage of the gas through it, may throw a drop or two through the gauze above, against the inner orifice of the jet tube, which would cause a sputtering in the flame.

The oil should be emptied out from the trough when the apparatus is laid by. Fig. 4. is a section of the trough and part of the reservoir, drawn on a large scale, in order to render its construction more distinct.

A.A.A. is the reservoir. B.B. a brass tube (the trough) closed at the bottom, and fixed air tight into the reservoir. C. is a small tube in the interior of the reservoir; its upper orifice is covered with fine wire gauze, and reaches nearly to the top of the reservoir; its lower orifice is inserted into the bottom of the trough; four holes are made from the trough into the tube, and open a communication to the gases in the reservoir: a circular flat valve, D, lined with oiled silk or leather, and moveable on a central pin, E, covers these holes, and prevents the passage of any thing from the trough into the reservoir, F, a fine wire gauze intersecting the trough. The head of the trough (*d.* fig. 1.) contains a small chamber, G. communicating by a fine tube with the interior of the trough, just below the orifice of which is a second piece of very fine wire gauze, M. The stop-cock, H, connects the head with the jet, having a circular motion by the ball and socket joint, I, to which various tubes, as K, may be adapted. The atmospheric air, which he compressed between his knees, with a force proportionate to the blast required. When empty, it was filled again by the mouth through a separate tube, furnished with a stop-cock and fitted to the bladder.

By these pretended improvements, motions more or less troublesome have been substituted for a slight exertion of the muscles of the cheeks, and their inventors have demonstrated by their very contrivances that they did not know how to use the blowpipe; they might as well have proposed to play on a wind instrument with a bladder. Our conclusion must be, that all apparatus of this kind is perfectly useless.

OF THE COMBUSTIBLE.

Every kind of flame, provided it be not too small, is calculated for experiments with the blow-

line at L marks the height to which the oil should rise in the trough. For further security, Mr. Newman informs me, that he puts several pieces, to the number of twenty or thirty, of very fine wire gauze, between the stop-cock, H, and the ball and socket joint, I, and the end of the reservoir nearest the syringe is made weaker than any other part, so that if an explosion *should happen* in the reservoir, it will yield in that part rather than in any other. With these precautions, the instrument may be considered (provided there be no fault in its construction, and every thing in good order), perfectly secure. C.

pipe, whether it be that of a tallow or wax candle, or of a lamp. Engeström and Bergman employed in preference common wax candles, furnished with a good cotton wick, which Bergman recommends to be bent, after snuffing, in the direction in which the flame is to be impelled. Candles however have this inconvenience, that the radiant heat from the substance under examination melts the tallow or wax, and occasions them to burn away too fast; and, besides, common candles do not always furnish sufficient heat. Gahn substituted at first for the single candle three small wax candles with thick wicks, which he placed together, but he afterwards rejected these for a lamp furnished with a large wick, and fed with olive oil. Lamps are certainly preferable to candles, though inconvenient in travelling, from the danger of the oil getting out. That, however, may easily be remedied by a brass cap made to screw over, or rather into the projecting piece (for the screw is best in the inside) through which the wick passes, and furnished with a washer of leather (previously soaked in melted wax), which, when the cap is screwed home, presses firmly on the rim of the projecting piece, and effectually prevents the escape of any oil from the lamp.[1] The best fuel for the lamp is olive oil.

[1] In point of form that represented at L (pl. I. fig. 10.) is the most convenient that I am acquainted with. It is made of copper, tinned both inside and out, and of the size drawn in the figure. C.

When we wish to heat a small matrass, or glass tube, a spirit lamp is preferable to the oil lamp. It should have a ground glass cap fitted to the neck of the lamp and covering the wick, to prevent the evaporation of the spirit when not in use.

OF THE BLAST AND FLAME.

The organs of respiration are not called into increased action in using the blowpipe; they could not keep up a continued blast, and such an effort would in time be injurious. It is the cheeks which perform the office of a pair of bellows: the mouth is filled with air, and by the contraction of the muscles of the cheeks it passes into the blowpipe. This operation, simple as it seems, is difficult at first, from the habit of exerting all the muscles concerned in respiration when we blow. It is a difficulty like that which a man experiences when he endeavours at the same time to turn his right arm and right leg in opposite directions. A little wearisome practice, therefore, is necessary to get over the custom of bringing the muscles of the chest into action with the muscles of the cheeks. The first thing to be attended to, is to keep the mouth full of air, during a pretty long alternation of inspirations and expirations; next, we must consider that there is a small opening between the lips, by which the air escapes, so that the cheeks would collapse by degrees, if the breath from the lungs were cut off from entering into the mouth. Now, to fill the vacuum which is thus formed, at the moment of expiration we admit a portion of air into the cavity of the mouth sufficient to renew the distention of the cheeks. Thus the air in the mouth is always in an equal state of compression, and escapes with uniform velocity through the little orifice. Such is the mechanism of the operation in which the art of using the blowpipe consists. We may add, that the current of air which escapes by the beak of the instrument, is commonly so minute, that it is not necessary to fill the cavity of the mouth at each expiration.[1] This operation, though somewhat difficult at first, soon becomes easy by practice, and at length is performed without occasioning the least distress to the respiration. The only inconvenience that remains, when once arrived at that point, is a lassitude in the muscles of the cheeks, arising, independently of want of practice, from the beginner's generally pressing the mouth piece of the blowpipe more strongly than is necessary between his lips, and not sufficiently economizing the blast.

Having accomplished the first object of keeping up a steady blast, the next is to produce a good

[1] In fewer words, the operator must breathe through his nostrils, and blow with his mouth by the mere compression of the cheeks. C.

heat, which requires some knowledge of flame, and of its different parts. If we attentively consider the flame of a candle, we may remark several unequal divisions in it, of which four may be distinguished. Plate II. fig. 5, represents the flame of a candle in its usual form. We see at its base a small part, *a, c*, of a dark blue colour, which becomes thinner as it gets farther from the wick, and disappears entirely where the external surface of the flame ascends perpendicularly. In the middle of the flame is a dark place, *a, d*, seen through its brilliant covering. This space encloses the gases which issue from the wick, which, not being yet in contact with the air, cannot undergo combustion. Round this space is the brilliant part of the flame, or the flame properly so called; and lastly, beyond this, we may perceive by attentive inspection the outer covering of all, *c, e*, slightly luminous, and whose greatest thickness corresponds with the summit of the brilliant flame. It is in this outer part that the combustion of the gases is completed, and the heat the most intense. If we introduce a fine platina or iron wire into the flame, we see that the point of the wire where the ignition is most vivid, is situated on the confines of the brilliant flame, and in the external covering. If the wire be very fine, its real diameter appears singularly magnified, and this apparent enlargement (which is a phenomenon of radiation of the same kind as that presented by the fixed stars, when we ascribe appreciable diameters to them), increases as we approach the upper boundary of the blue flame, so that this zone of transition, where the air, as yet retaining its full dose of oxygen, begins to meet the flame, is the place of the maximum of heat. This being granted, if we direct a current of air by the blowpipe into the middle of the flame (pl. II. fig. 6), a long narrow blue flame, *a, c,* appears before the opening in the beak, which is the same as *a, c,* in fig. 5, but its relative position is changed; instead of surrounding the flame it is now concentrated within it, where it forms a small cylinder. Toward the anterior extremity of this blue flame is the place of greatest heat, just as in the flame not acted on by the blowpipe. But whilst in the latter this place had the form of a zone, or circumference of a circle, it is now reduced to a point incomparably hotter, and capable of fusing or volatilizing substances on which the flame in its common state has no sensible action. This enormous increase of temperature arises from the blowpipe throwing a condensed mass of air, which before only touched the surface of the flame, and spread itself freely about every part of it, on a small space situated in the middle. The change effected is somewhat the same, as if the flame had been turned inside out. On the other hand, the remaining portion of the bright flame which here surrounds the blue, prevents the loss of the heat produced.[1]

[1] We are indebted to Sir Humphry Davy for correct notions respecting flame. In the course of the experiments

Long practice is necessary to distinguish with certainty the maximum of heat, seeing that differ-

which led to the invention of his safety lamp (that glorious triumph of science over destruction), he successfully investigated the nature and properties of flame. He showed that flame (which he defines to be gaseous matter, heated to such a degree as to be luminous), in all cases, must be " considered as the combustion of an *explosive mixture* of inflammable gas, or vapour, and air, for it cannot be regarded as a mere combustion at the surface of contact of the inflammable matter; and the fact is proved by holding a taper, or a piece of burning phosphorus, within a large flame made by the combustion of alcohol; the flame of the candle, or of the phosphorus, will appear in the centre of the other flame, proving that there is oxygen even in its interior part." (Davy on the Safety Lamp, &c. p. 46.) He showed that a very high temperature (but differing in degree for different explosive mixtures, in some inverse ratio of their inflammabilities, or of the heat they evolve in combustion) is necessary for these successive explosions to take place. " If a piece of wire gauze sieve is held over the flame of a lamp, or of coal gas, it prevents the flame from passing it;" " the air passing through is found very hot, for it will convert paper into charcoal; and it is an explosive mixture, for it will inflame if a lighted taper be presented to it, but it is cooled below the explosive point by passing through wires even red hot, and by being mixed with a considerable quantity of air comparatively cold." (P. 47.)

He showed that the heating effect of explosive mixtures depends on their containing such relative proportions of inflammable gas and oxygen, as to ensure the perfect combustion of all the inflammable matter whilst in its gaseous state, but that a high degree of illuminating power results from an opposite cause, namely, the "decomposition of part of the gas towards the interior of the flame where the air is in ent bodies have different modes of ignition, and that we are easily deceived by the light which they

smallest quantity, and the deposition of *solid charcoal*, which, first by its *ignition*, and then by its *combustion*, increases in a high degree the intensity of the light." (P. 50.) Hence the feeble pale coloured light when a safety lamp is burnt in a very explosive mixture of coal gas and air, and the brilliancy of the flame of a stream of coal gas burnt in the atmosphere, as he proved by the following experiments. "I held a piece of wire gauze, of about 900 apertures to the square inch, over a stream of coal gas issuing from a small pipe, and inflamed the gas above the wire gauze, which was almost in contact with the orifice of the pipe, when it burned with its usual bright light. On raising the wire gauze, so as to cause the gas to be mixed with more air before it inflamed, the light became feebler; and, at a certain distance, the flame assumed the precise character of that of an explosive mixture burning within the lamp; but though the light was so feeble in this last case, the heat was greater than when the light was much more vivid, and a piece of wire of platinum held in this feeble blue flame became instantly white hot.

"On reversing the experiment by inflaming a stream of coal gas, and passing a piece of wire gauze gradually from the summit of the flame to the orifice of the pipe, the result was still more instructive, for it was found that the apex of the flame, intercepted by the wire gauze, afforded no solid charcoal; but in passing it downwards, solid charcoal was given off in considerable quantities, and prevented from burning by the cooling agency of the wire gauze; and at the bottom of the flame, where the gas burnt blue in its immediate contact with the atmosphere, charcoal ceased to be deposited in visible quantities.

"This principle of the increase of the brilliancy and density of flame by the production and ignition of solid matter, appears to admit of many applications.

"It explains readily the appearances of the different parts

emit. To attain this maximum we must neither blow too strongly nor too gently: in the first case, the

of the flames of burning bodies, and of flame urged by the blowpipe; the point of the inner blue flame where the heat is greatest, is the point where the whole of the charcoal is burnt in its gaseous combinations without previous deposition." (P. 50, 51.)

At p. 54. Sir Humphry adds, "whenever a flame is remarkably brilliant and dense, it may be always concluded that some solid matter is produced in it: on the contrary, when a flame is extremely feeble and transparent, it may be inferred that no solid matter is formed.

"The heat of flames may be actually diminished by increasing their light (at least the heat communicable to other matter), and *vice versa*. The flame from combustion which produces the most intense heat amongst those I have examined, is that of a mixture of oxygen and hydrogen in slight excess, compressed in a blowpipe apparatus, and inflamed from a tube having a very small aperture. This flame is hardly visible in bright day-light, yet it instantly fuses very refractory bodies; and the light from solid matters ignited in it, is so vivid as to be painful to the eye." (P. 55.)

"The form of the flame" (of a common candle for instance) "is conical, because the greatest heat is in the centre of the explosive mixture. In looking stedfastly at flame, the part where the combustible matter is volatilized is seen, and it appears dark, contrasted with the part in which it begins to burn, that is, where it is so mixed with air as to become explosive. The heat diminishes towards the top of the flame, because in this part the quantity of oxygen is least. When the wick increases to a considerable size from collecting charcoal, it cools the flame by radiation, and prevents a proper quantity of air from mixing with its central part; in consequence, the charcoal thrown off from the top of the flame is only red hot, and the greater part of it escapes unconsumed." (P. 102).

heat is carried off as soon as produced by the impetuosity of the current of air,[1] and besides a part of the air escapes without assisting the combustion; in the second, sufficient air is not supplied in a given time. A very high temperature is necessary whether we wish to try the fusibility of a body, or whether we have to reduce certain metallic oxides, which part with their oxygen with difficulty, as the oxides of iron and tin. But our pyrognostic

The internal structure of flame has been neatly exhibited by Mr. Oswald Sym, by dissecting it perpendicularly to its axis, by means of wire gauze held horizontally across it. The inference he draws from his experiment is, that the entire flame is a hollow substance, that "the actual combustion is confined to the surface, and that the internal part is filled with volatilized wax. In short, the flame of a candle is an elliptical bubble." Some of Mr. Sym's conclusions have been called in question by Mr. Porrett, particularly his assertion that "flame is an opake substance;" but this note is already so long, that I can only refer the reader who may wish to pursue the matter farther, to those gentlemen's papers in the Annals of Philosophy, the first in vol. viii. p. 321, the second in vol. ix. p. 337. C.

[1] Perhaps the explanation may be more satisfactory in the words of Sir Humphry Davy. "A large quantity of cold air thrown upon a small flame lowers its heat beyond the explosive point, and in extinguishing a flame by blowing upon it the effect is probably principally produced by the same cause." (Safety Lamp, p. 46.)

The injurious effect of blowing too strongly, may, I conceive, rather be owing to the extinction of a portion of the flame, in the manner above mentioned, than to the heat being carried off as soon as produced, by the impetuosity of the current. C.

operations are not confined to obtaining the highest possible temperature; other phenomena must be produced which require a less intense heat. These are, oxidation and reduction, which are both easily affected, although diametrically the reverse of one another.

Oxidation ensues when we heat the subject under trial before the extreme point of the flame, where all the combustible particles are soon saturated with oxygen; the farther we recede from the flame, the better the oxidation is effected (provided we can keep up sufficient heat); too great a heat often produces a contrary effect, especially when the assay is supported by charcoal. Oxidation goes on most actively at an incipient red heat. The opening in the beak of the blowpipe must be larger for this kind of operation than in other cases.

For *reduction*, a fine beak must be employed, and it must not be inserted too far into the flame of the lamp; by this means we obtain a more brilliant flame, the result of an imperfect combustion, whose particles, as yet unconsumed, carry off the oxygen from the subject of experiment, which may be considered as being heated in a species of inflammable gas. If in this operation the assay become covered with soot, it is a proof that the flame is too smoky, which considerably diminishes the effect of the blast. Formerly, the blue flame was considered as the proper one for the reduction of oxides, but this idea is erroneous; it is in reality the brilliant part of the flame which produces deoxidation: it must be directed on the assay so as to surround it equally on all sides, and defend it from the contact of the air.

I repeat, it is the combustible atmosphere, in which the assay is immersed, that most powerfully promotes its reduction; for that produced by the charcoal, at its point of contact with the oxide, takes place as well in the exterior as in the interior flame.

The most important point in pyrognostic assays is the power, easily acquired, of producing at will either oxidation or reduction. Oxidation is so easy, that one need merely be told how it is to be done, to be able to do it; but reduction requires more practice, and a certain knowledge of the different modes of conflagration. A very advantageous mode of practice, in order to acquire the art of making a good reducing flame, is to fuse a small grain of tin, and raise it to a reddish white heat on a piece of charcoal, so that its surface may always retain its metallic brilliancy. Tin has so great a tendency to oxidation, that the moment the flame begins to become an oxidating one, it is converted into an oxide of tin, which covers the metal with an infusible crust. We must begin by operating on a very small grain, and gradually proceed to larger and larger. The greater the quantity of tin that he can thus keep in the metallic state, at a high temperature, the more expert is the operator in his art.[1]

[1] If a red hot grain of metallic tin be thrown on a paper

OF THE SUPPORT.

Charcoal.—The substance to be examined by the blowpipe must necessarily rest on a solid body, or be fixed in some manner in a steady position. The best support is charcoal. The wood of a sound well grown pine tree, or light woods in general are preferable. The charcoal of the fir tree is liable to crackle and scintillate, and to scatter the assay.[1] The charcoal of hard compact woods gives so much ash, and that ash is sometimes so ferruginous, that it should never be employed but for want of better; for this reason, beech and oak are not fitted for the purpose. I have never yet had an opportunity of trying box-wood charcoal. Gahn always conceived that it would be the best of all for blowpipe experiments, but he seldom had the choice of any other than our pine-wood coal. Of the various kinds which I have tried in countries where those above mentioned are not found, it appeared to me, that the charcoal of white willow, or of the willow species in general, was the best; but I still prefer that from the mature pine, cleft in the direction of the grain,[2] which may be divided by the saw into long paralellopipedons, and cleaned by simply

tray, it will divide into several smaller grains, which skip about the paper and burn with a very vivid light. B.

[1] I shall for brevity generally use this term to denote the subject of experiment. C.

[2] "Débité à droit fil."

blowing off the dust. In order to fix the flux to a point on the surface of the support, one of the ends perpendicular to the layers of the wood is to be chosen for its receptacle; if placed on the section parallel to the layers, it would spread over the surface. Although the space between the woody layers may consume faster than the layers themselves, there is this advantage attending it, that the assay rests in that case merely on their summits, and thus often has only two or three points of contact with the charcoal.

It is almost needless to observe, that the charcoal we use must be well burnt: that which splits, smokes, or burns with flame, is unfit for the purpose. Gahn had an idea that the charcoal which sometimes descends in our graduated furnaces (fourneaux graduées) down to the air hole, without consuming, would be better than any other for blowpipe experiments, because, having lost a portion of its combustibility, it ought not to burn away so fast. I, in consequence, collected a quantity of this charcoal. It was heavier, and more compact than common charcoal, and at the same time very difficult to burn; but I was surprised to find that I could not obtain so high a heat with it as with the very combustible coal I usually employ. I at first supposed this difference to arise from the radiant heat (which in common charcoal flows from the ignited part near the assay), being essential towards raising the temperature; but I soon discovered my mistake, when on taking hold of my new support, at a part very remote from the heated point, I found it so hot that I was obliged to let it go. Charcoal, therefore, in acquiring hardness, becomes a good conductor of heat, and the denser it is the better it conducts it: this, perhaps, is the reason also why it burns so slowly; hence we see that it is unfitting for experiments with the blowpipe.[1]

Platina.—In those cases where the reducing effect of charcoal would be injurious, a support of platina is employed, sometimes in the form of a spoon, sometimes in that of thin foil or wire.

Platina Spoons.—Gold and silver spoons were formerly used when a mineral was to be fused with soda, but being liable to be melted, platina spoons have been substituted for them. These, however, are become quite superfluous in experiments with the blowpipe, since it has been found that mineral substances may be heated with soda or charcoal, better than on any other kind of support. Besides, the size of the spoon[2] prevents our obtaining the

---

[1] The best charcoal that I know for the use of the blowpipe, is that which is made from *alder*, and is employed by the English manufacturers in making the coarser kinds of gunpowder. It is light, but sufficiently compact, extremely even in its texture, and, when well burnt, neither smokes, splits, nor scintillates. Straight pieces, and free from knots, should be selected. Box-wood charcoal is not eligible; its density renders it too good a conductor of heat, and it is very apt to split; besides, it is not easy to obtain it in pieces of sufficient size. C.

[2] By making the spoon of very thin foil, this objection

Fig. 4.

Fig. 5.

Fig. 6.

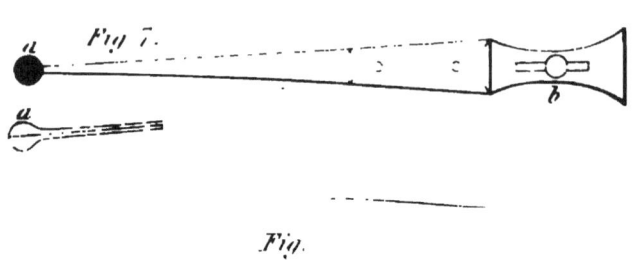

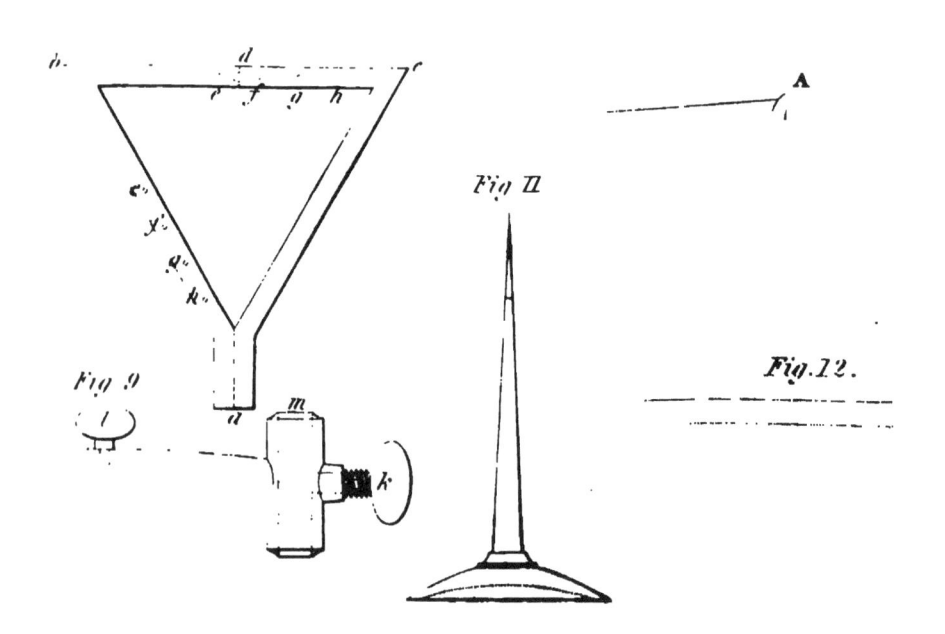

Fig. 7.

Fig.

Fig. 11.

Fig. 9.

Fig. 12.

Published by Baldwin, Cradock & Joy, March 1822.

very high temperatures, which are often necessary.

Platina foil.—Wollaston substituted very thin platina foil, cut into slips about two inches long and half an inch wide, in the place of spoons. This foil, from its thinness, may be intensely heated, and when we wish to heat and oxidate at the same time, the flame is to be directed against the lower surface. Platina is so bad a conductor of heat, that the slip of foil may be held in the fingers by one end, whilst it is red hot at the other. Substances in the metallic state, or those oxides which are reducible *per se* before the blowpipe, must not be supported on platina foil; for the platina will alloy with them, and fuze into holes.

Platina Wire.—Gahn, who knew the inutility of spoons, but was not acquainted with the use of platina foil, employed a platina wire two inches and a half long, and bent at one end into a hook, (pl. III, fig. 1) which serves as the support in the manner following. Having moistened the hook with the tongue, it is to be dipped into the flux, a portion of which will adhere to it; this is to be fused by the lamp into a globule, which congeals and adheres to the curvature. The assay must then be moistened, to make it adhere to the flux, which is now solid, and the whole heated together. We thus obtain an insulated mass, which may be

---

might perhaps be done away, but still its form is objectionable and disadvantageous. C.

water, or any other volatile incombustible contained in a mineral; or if the mineral decrepitate violently, it is best heated in a matrass or flask, (pl. III, fig. 3) whose body is sufficiently large to allow the air to circulate, and the volatile substances to rise. But if we have combustible substances, as sulphur, arsenic, &c. to separate from the assay by sublimation, the body of the flask must be small, since the renewal of the air in a matrass of the former kind, might occasion the combustion of the disengaged substances.[1] Engeström heated decrepitating substances in a hole made in a piece of charcoal, and covered with another piece, leaving a small opening to introduce the flame. Bergman used a glass tube, or a spoon with a cover. Wollaston wraps the substance in platina foil.

As the matrass (fig. 3) is seldom so much heated as to be injured, it may be used repeatedly; but the glass tubes, especially the open ones, cannot be heated a second time, without danger of their breaking at the place where the assay rested, and they were exposed to a high temperature; but that portion may be cut off with a file, the interior cleaned, and the end closed again if necessary, and then they may serve for several experiments. It is well, however, to have a good supply of each sort, as they are continually in request.

---

[1] A glass tube of small diameter, and closed at one end by the lamp, is best adapted to this operation. C.

ADDITIONAL INSTRUMENTS.

Forceps are used to hold a small lamina of a substance, when we wish to try its fusibility. Their extremities, at least, should be made of platina. Pl. III, fig. 4 and 5, represent the form of a pair on the best construction, seen in two directions; *ab, ab,* are two thin plates of steel, each having a piece of platina rivetted on its extremity, *b.* These plates are fastened in the middle to the same iron plate, *e, e,* so as to form a double pair of forceps, steel at one end, and platina at the other. An interval is left between the plates at the steel end, but the platina extremities are made to shut close together, by the spring of the steel to which they are rivetted. To open them, we press the fore finger and thumb against the two buttons, *d, d,* each of which is fixed in one plate, and passes through the other. The platina pieces must be strong enough at *b,* to resist the pressure of the steel spring, diminishing in breath and thickness from *b* to *c,* that they may not carry off the heat too much from the assay. The steel ends must be hardened at *a,* that they may not be *battered* when used to detach a particle for fusion from the mineral to be examined. These forceps were brought from France, and perfectly answer their purpose.

Those used for the same purpose in England

have a different and less convenient form. Pl. III, fig. 6, is a side view of one of these instruments. The branches, *ab*, are of brass, fixed together at *a*; they terminate like the former in platina points rivetted on; they stand open, but may be closed by a button, *d*, with two heads, which slides backwards and forwards in a slit cut in the branches. When the button is pushed from *a*, towards *c*, the branches approach each other, and their points touch when it is placed against the pieces, *e, e,* which are nothing else but two pieces of wood fixed to the branches, and by which the forceps are held. Brass is so good a conductor of heat, that without this precaution we should be in danger of burning our fingers after the blast had been kept up for a few moments on the substance contained between the points.[1]

Pl. III, fig. 8, is a pair of forceps for large pur-

---

[1] Fig. 7 is a pair of platina forceps, the extremities of which are excavated similar to an ear pick, *a*; and, when shut, will contain any small gem or substance liable to decrepitate during the application of heat. They are provided with a slider, *b*, that secures them from opening during the process, and will be found extremely useful in examining bodies which are liable to dispersion when heat is applied.

For fig. 7, and the preceding description of it, I am obliged to my friend, Mr. Pepys, by whom this very serviceable little instrument was invented. Decrepitation is a momentary phenomenon; it ceases when the assay has been once heated red; after that it may be treated without any envelope. C.

poses. The button, *d, d,* is furnished with a steel spring behind; to prevent its sliding back when the cheeks are opened wide.

Cutting pliers, made thick and strong in the cutting part, are useful for detaching small portions from a specimen, without injuring it.

A *pair of pliers,* terminating at one end in a point, are useful in trimming the lamp.

Hammers.—Two of hardened steel are necessary. The face of one should be round and polished; it is used to flatten the grains of reduced metal: its other end must have the form of a narrow cone, with an obtuse summit, and serves to detach portions of the mineral to be examined, when it is required to confine the blow to a small part of the surface. The face of the other hammer should be square, and its edges sharp; its opposite end is formed like, and serves the purpose of, a chissel. This hammer is useful for chipping off small portions from a specimen for examination. All its edges should be kept sharp.

Anvil.—A paralellopipedon of steel of about 3 inches long, 1 inch thick, 3-4ths of an inch wide, and polished on all its faces, is most convenient for this purpose. It is used to crush pieces of minerals, which are to be wrapped in paper and laid on the anvil, and struck with the hammer. The paper prevents the dispersion of the fragments. When we would try the malleability of a grain of reduced metal, it is to be treated in the same manner.

Bergman's iron ring, which he placed on the

anvil to prevent the fragments from being scattered about, is unnecessary, and, besides, does not answer the purpose.

Knife.—A knife of hard steel of the size of a penknife, with a sharp but not too fine edge and point, is useful for trying the hardness of bodies, which is estimated by the greater or less resistance they oppose to it. Its point also, previously moistened by the mouth, serves to take up portions of the fluxes, and, if necessary, to knead them with the pulverized mineral in the palm of the left hand. In short, the knife is one of the most indispensable instruments used with the blowpipe.

Files.—A triangular, a flat, and a round, or half round file are useful.

A small agate mortar and pestle. The smaller the better. That which I use is scarcely two inches in diameter, and half an inch high on the outside. Its cavity is 5-16ths of an inch less, in breadth and depth.[1]

It is desirable that the bottom of the mortar be somewhat transparent; but it should be quite free from cracks and crevices, in which the pulverised substances would lodge.

---

[1] Gahn having once lost the pestle of a similar mortar, took a button of calcedony of suitable diameter, and fixed it with sealing wax to a cork. This new pestle was the only one he ever used afterwards. I have been obliged to have recourse to the same expedient, and have thought it right to mention it in this place, for the use of those who may fall into a similar predicament. B.

A small piece of pummice stone should be kept at hand, to remove the traces sometimes left on the surface of the mortar by metalliferous substances.

A *Miscroscope* with one or two plano-convex lenses of different powers, which form Dr. Wollaston has shown is best adapted to enlarge the field of distinct vision. In the sort of experiments we are treating of, the miscroscope is often indispensable to ascertain the result, and we must be cautious how we decide as to colour, before we have examined the object by its means, for the light reflected by the charcoal on small globules of glass often produces false appearances which the microscope corrects.

A box with compartments to hold the fluxes, and a tray of sheet iron not tinned, lined with white paper, to catch the particles that may fall from the support during an experiment, are useful.

A glass tube with a fine orifice, passing through a cork fitted to the neck of a phial half full of water, is convenient for washing the charcoal powder from the grains of reduced metal, in certain experiments.[1]

---

[1] This is a neat contrivance of Dr. Wollaston's for dropping water. Berzelius calls it the "*fountain of compression,*" and describes the mode of using it, thus. " The cork is fixed in the neck of a flask half full of water, which is turned upside down, and into which *we blow, so as to compress the air within.* This air, by afterwards dilating, produces a fountain;" which he might have added, will play in the face of the operator,

A polished blood stone is useful to rub the metallic substances on, that have been reduced but not fused, in order to ascertain if they be possessed of metallic brilliancy, &c.

Small porcelain capsules[1] are useful to hold the fragments that have been, or are to be examined, &c.

Order in the arrangement of the various instruments is very essential, so that the operator may in a moment lay his hand on whatever he may want.[2]

The following apparatus is not particularly connected with the uses of the blowpipe, but is convenient. Pl. III. fig. 9, *i*, *k*, is a small vice sliding on an upright brass rod, furnished with a foot (not shown in the plate), to which it is fixed by the screw, *k*; the opening, *i*, receives the triangle, *a*, *b*, *c*,

as soon as he removes the bottle from his mouth. The dilatation of the air by the heat of the hand, is all the expansive power necessary to expel the water, and I believe the only one intended to be applied. It is convenient to have the tube bent nearly at a right angle, about an inch from the orifice. C.

[1] Or watch glasses. C.

[2] Here follows a long detailed description of a table, with a drawer at each side, and four in front, divided into moveable compartments of tinned iron, to hold the various instruments, &c. not forgetting *a hook with a towel fixed to the right leg of the table!* Next comes an equally elaborate description of a red morocco case to hold a travelling blowpipe apparatus. These things are all very useful, but I cannot agree with my author, that a particular description of them is necessary. I have, therefore, omitted them. C.

and is tightened by the screw, *l*. At *d*, is a joint to fold the triangle together for the convenience of travelling, &c.; *e, f, g, h,* on the side *a, b,* are small holes, each intended to receive one end of a steel wire, bent at a right angle, whilst the other end, similarly bent, fits into the corresponding hole on the side *b, c*. The large triangle, *a, b, c,* may thus be divided into four smaller ones, according to the diameter of the capsule or crucible intended to be supported by it. The triangle is made of strong iron plate, and the length of its side is about $2\frac{1}{4}$ inches.

A magnetic needle, A, and Hauy's Electrical needle, B, (pl. III, fig. 10,) to which the pivot (fig. 11.) serves as a common support, are necessary; also a small bar magnet, to produce an equilibrium between the mutual action of the magnet and the magnetic needle, and that of the needle and the earth, after the manner pointed out by M. Hauy.[1]

[1] This experiment for rendering very weak magnetic attractions perceptible, is performed as follows. The magnetic needle is suspended on its point, and when it has taken its position in the magnetic meridian, the magnet is placed at a certain distance from it, the north pole, for instance, of the magnet, being opposed to the north pole of the needle. The magnet is then gradually approached towards the needle, till, from the mutual repulsion of the similar poles, the needle has taken a direction perpendicular to that which it had at first. The repulsion of the magnet is now in equilibrium with the magnetism of the earth, and the least force acting on the needle will overcome it. By this arrangement, very small magnetic attractions may be rendered evident, which

A cylindrical case open at both ends. (Pl. III, fig. 12.) One half of the cylinder is occupied by a foot or stand made of a glass tube, filled with sealing wax, and terminated by a small steel point, on which the electrical needle, B, (fig. 10,) oscillates, when it is required to insulate the stand. The other end of the case contains a conical stand, also made of sealing wax, to which is fastened the hair of a cat. This electroscope, invented by Hauy, is the most sensible known, and renders all others almost unnecessary in physico-mineralogical researches. The hair is negatively excited,[1] by drawing it through the fingers, after which it is attracted or repelled by any electrical body to which it is presented, according as that body is positively or negatively electrified. If the hair have not been excited, it is attracted alike by all electrical bodies, and discovers electricity of too feeble intensity to have any influence on the electrical needle.

To these articles may be added a pair of small

---

would have no action on the needle in its ordinary position. In this experiment, care must be taken not to communicate, in handling it, any electricity to the mineral examined, since it would act on the needle as a magnetic power. B.

[1] " Ou frotte le poil entre ses doigts pour y susciter l'electricité *négative*." I think this is a mistake. I stuck a hair from a cat's back to the end of a stick of red sealing wax, and excited it by drawing it through my fingers. It was attracted by excited sealing wax, and repelled by an excited glass tube. The same effect took place when I excited the hair by silk. Its electricity must, therefore, be positive. C.

scissars, a pair of small tongs to hold crucibles, &c. when heated by the spirit lamp; a touch stone, and some assay needles, made of alloys of gold of different standards, for trying the purity of gold.

OF THE REAGENTS, AND THEIR USE.

Cronstedt commonly employed only three reagents,—subcarbonate of soda, borate of soda, and the double salt formed of phosphate of soda, and phosphate of ammonia. I shall henceforth, for the sake of brevity, call these simply soda, borax, and salt of phosphorus.

These reagents are still employed; and in the multitude of substances which have been tried since the time of Cronstedt, there is not one preferable to either of the above in their respective uses. It is singular, that in the infancy of the art, the best should have been hit upon. To these, which are the principal, and always in request, others have since been added, not so frequently wanted indeed, but which it is necessary to have at hand for particular cases. I shall treat, in succession, of the different salts employed, and explain the object to which each is applicable, and the method of using it.

Soda.—(Carbonate of Soda.) We may take indifferently, either the carbonate, or the bicarbonate, but both must be perfectly pure, and especially, free from sulphuric acid. The best method of obtaining the bicarbonate, is, to impregnate a concentrated solution of purified soda, such as is found at the druggists, with carbonic acid gas. In this operation the bicarbonate precipitates in the form of small crystallized grains, which are to be twice washed in cold water and dried. The salt, either previously ignited or not, is reduced to fine powder. If not ignited, it is more bulky, but there is this inconvenience from igniting it, that if the knife with which we take up portions of it be wet, the moisture gradually communicates to the whole mass, and causes it to collect into coarse lumps.

Two principal objects are connected with the use of soda; 1st, to ascertain if bodies combined with it be fusible or infusible; 2dly, to assist the reduction of metallic oxides. In both these respects it is one of the most indispensable reagents.

(*a.*) *Of the fusion of bodies by soda.* A very large number of bodies have indeed the property of combining with soda at a high temperature, but most of these combinations are infusible. It is only with acids, and a small number of metallic oxides, including silica, that it forms fusible compounds, which for the most part are absorbed by the charcoal.

Of these compounds, the glass formed with silica, or siliceous minerals, of which I shall speak more fully when we treat of that oxide and the mode of examining the silicates, occurs most frequently. With regard to the use of soda, many minutiæ are to be observed. We take up the necessary quantity with the point of a small knife, previously moistened by the mouth to make the powder adhere to it. It is then applied to the palm of the left hand, and, if necessary, moistened again, so as to form it into a coherent mass. If the substance to be tried be pulverulent, it must be kneaded in the hand with the soda, but if it be in the form of a spangle or grain, the soda is applied to its surface; it is then heated on charcoal, at first to dryness, and afterwards till it fuses. It commonly happens, that the soda is absorbed by the charcoal the instant after it is melted; but this does not prevent its effect on the assay; for if it be fusible with the soda, it soon after pumps it up again, (la repompe), a continued effervescence appears at its surface, its edges become round, and the whole is transformed into a fluid glass globule. If the assay be infusible in soda, but decomposable by it, it is perceived to swell up by degrees, and change its appearance without fusing. Ere we pronounce on the infusibility of any substance with soda before the blowpipe, we must always try a mixture of the flux, with the substance *reduced to powder.* If we take too little soda, a part of the matter remains solid, and the rest forms a transparent vitreous covering around it; if we take too much, the glass globule becomes opaque on cooling.

The assay may perhaps contain some substance, which, being infusible in the glass of soda, destroys its transparency; in that event, to avoid being deceived with respect to the nature of the glass, we must in the two before-mentioned cases, add a fresh portion of the matter supposed to be wanting, and thus endeavour to obtain a clear glass globule. In general, it is best to employ the soda in small successive doses, observing the changes produced by the different proportions of the flux. It sometimes happens, that the glass becomes coloured at the moment it begins to cool, and ends by assuming a tinge, which may vary from yellow to deep hyacinth red; occasionally it even becomes opaque and yellowish brown. These appearances occur if the soda contain sulphuric acid, or only sulphur, and the discoloration is the consequence of the hepar, or liver of sulphur, formed by the reducing action of the charcoal. If it always ensue with the same soda, it is a proof that it contains sulphate of soda, and must be rejected,[1] otherwise it is the assay which contains sulphur, or sulphuric acid.

Cronstedt and Bergman directed minerals to be fused with soda in a metal spoon, because they supposed the absorption of the soda by the charcoal would prevent its action on the assay. Gahn never employed a spoon in this case, for the fused mass spreads over metallic surfaces, whilst on charcoal it assumes a globular form, in which state we

[1] It is best to ascertain that the soda do not contain any sulphate before it is used. For this purpose, dissolve a small portion in distilled water; and, to the clear solution, add a few drops of muriate of baryta. If the precipitate be not wholly soluble in muriatic acid, the soda contains some sulphate. C.

can more accurately judge of its colour and transparency.

(*b.*) *Reduction of metallic oxides.*—This mode of assaying, by which we often discover portions of reducible metal so minute as to escape detection by the best analyses made in the moist way, is, in my opinion, the most important discovery that Gahn ever made in the science of the blowpipe.

If we place a small piece of native or artificial oxide of tin on charcoal, a skilful operator will be able, with some exertion, to produce from it a small grain of metallic tin; but if a little soda be added, the reduction is easily effected, and so completely, that, if the oxide be pure, it is wholly converted into reguline tin. Soda therefore manifestly favours the reduction; but we are ignorant of its precise mode of action. When the soda surrounds the substance to be reduced, we may imagine that this reagent, which, by its interposition between the assay and the charcoal, prevents rather than facilitates their contact, may itself be reduced in a certain degree by the action of the blowpipe, the consequence of which is the reduction of the oxide imbedded in the soda, by the base of the soda or its suboxide. If the metallic oxide contain a foreign body, incapable of being reduced, the reduction of the former often becomes more difficult in consequence; but if we add a little borax, this reagent will co-operate with the soda in dissolving the foreign substance, and the reduction will be speedily effected. This experiment is easily performed, and the nature of the metal the more readily ascertained, from our generally knowing by previous trials, that of the oxide operated on; hence the reduction is only a confirmation of a previous conclusion. Bergman has described this process.

Suppose now, that the metallic oxide be mixed with a comparatively large quantity of irreducible substances, and that its presence be either unknown, or impossible to be ascertained by other similar experiments; how are we to discover whether the assay contain a reducible metal, and if it does, how are we to prove it? Gahn has given a simple solution of this problem.

The assay, being reduced to powder, is to be kneaded in the hollow of the left hand with moistened soda, the mixture placed on charcoal and exposed to a good reducing heat; then a little more soda is to be added, and the blast renewed. As long as any particle of the assay remains on the surface of the charcoal, soda is to be added in small doses, and the blast continued till the whole mass is absorbed by the charcoal. The first portions of soda serve to collect the metallic particles, which were dispersed through the whole matter of the assay, and the final absorption of the latter completes the reduction of whatever till then had retained the state of oxide. This done, the charcoal is to be extinguished with two drops of water, and the portion containing the absorbed matter being removed with a knife, is to be reduced to very fine powder in an agate mortar.

This powder is then to be washed with water to free it from the charcoal which forms its lightest part, for which purpose we use the little fountain of compression.[1] The grinding and washing (the latter requires particular attention) are to be repeated as often as necessary, till the whole of the charcoal is got rid of. If the assay contained no metallic substance, nothing remains in the mortar after the last washing. But if it contained any, however small a portion of reducible metal, it will be found at the bottom of the mortar in the form of little brilliant leaves, if fusible and malleable; or as a powder, if brittle and not fused. In either case we perceive on the bottom of the mortar metallic traces, arising from the friction of the particles of metal against the agate, provided the quantity of metal contained in the specimen was not too small. The flattening of malleable metals transforms an imperceptible particle into a round shining disc of sensible diameter. Thus we may detect by the blowpipe in an assay of the usual size, $\frac{1}{4}$ per cent of tin, and even the slightest traces of copper.

We must endeavour in these experiments, 1st, to produce as strong a heat as possible, by directing the reducing flame, so as completely to cover the surface of the fused mass; 2dly, to leave nothing in the charcoal, and to lose none of the assay when we collect it; for though the particles

---

[1] See note, page 41. C.

of the metal may be fused into a globule in some part of the mass, we know not where the globule is; 3rdly, to grind the carbonaceous mass sufficiently; 4thly, to decant very gently, so that only the lightest and finest particles may be carried off by the water; 5thly, not to attempt to ascertain the result till the whole of the charcoal be got rid of; since the smallest remaining portion might be sufficient to conceal the little metallic spangles, and the particles of the charcoal themselves, seen in a certain light, display a metallic lustre, which may deceive an inexperienced eye: and, lastly, we must not be content with examining the result with the naked eye, but inspect it carefully with the microscope, if it be only to satisfy ourselves that we have judged accurately concerning it. When the bottom of the mortar, by long use, has become uneven, the small cavities which it contains are sometimes filled with air, and form bubbles under water, whose convexity reflects light like a polished metallic surface. The nature of the reflecting surface is easily detected by turning the bottom of the mortar to the light; if it be a bubble, the light will be seen through the spot it occupies, but if it be a spangle of metal, the spot will be opaque.

The metals reducible by this process (besides the noble metals), are molybdenum, tungsten, antimony, tellurium, bismuth, tin, lead, copper, nickel, cobalt and iron. Of these, antimony, bismuth, and tellurium are easily volatilized by a strong reducing flame. Selenium, arsenic, cadmium, zinc, and mercury, volatilize so completely, that they cannot be collected without a little subliming apparatus.

The reduction is always effected at the first trial, if the assay contain from 8 to 10 per cent. of metal. But as its proportion diminishes, more attention and practice are necessary to keep it in the mortar and ascertain its nature. A good method of practising oneself in this kind of operation, is to select a substance containing copper, and make several experiments of reduction with it, mixing it every time with a larger proportion of matter containing no copper; thus the metallic standard diminishes with every fresh experiment, and when we have arrived at such a proportion that we cannot effect the reduction of the copper by a first operation, we must repeat the trial till we succeed in developing it. By exercising oneself in different trials of this kind, we soon become expert in such operations, which only require a little address and practice.

If the assay contain several metals, the reduction of these oxides is commonly made *in globo*,[1] and we obtain for our regulus a metallic alloy. A few reduce separately, as copper and iron, which give a regulus of each metal,—copper and zinc, the former gives a regulus of copper, the latter sublimes. But when the result of the operation is an alloy, in order to find what metals it is composed of, we must have recourse to particular effects pro-

---

[1] So in the text. C.

duced by other fluxes. Further on I shall describe the distinguishing characters of each metallic oxide.

For want of soda, we may employ potassa in all the preceding instances; but soda is preferable, first, because its carbonate is not deliquescent like that of potassa, and, secondly, the capacity of saturation of soda is to that of potassa, (all circumstances being equal) as 25·58 : 16·95; that is to say, in the ratio of the quantity of oxygen they contain.

2. *Borax.*—The borax of commerce should be dissolved and crystallized afresh, before it is used for experiments with the blowpipe. Gahn used often to remark, that in fusing borax of commerce with soda on charcoal, till the two salts are absorbed, a white metal is frequently obtained from the mass, which seems to be derived from the vessels in which the borax is refined. This does not happen with re-crystallized borax.

The borax is kept either in small grains, of the size required for experiment, or in powder, in which case it is taken up like the soda, with the moistened point of the knife. Some operators fuse it to get rid of its water of crystallization, and thus avoid the intumescence which preceeds the fusion of the crystal containing water, on the charcoal. This were a very good precaution, but that the borax soon recovers its water of crystallization, and swells up before the blowpipe as at first, though in a rather less degree. I always use my borax un-calcined, for the intumescence is a slight inconvenience, and it is easy to melt the mass into a globule.

Borax is used to effect the solution or fusion of a great number of substances. According as the matter to be dissolved is pulverulent or granular, it is spread over the borax at the instant of its swelling up, or fixed on the fused globule by being moistened. In general, we begin by attempting the solution of a *spangle*, because, if the assay be in the form of a *powder* in experiments of this kind, we cannot distinguish accurately the parts of the assay not attacked by the flux, from certain insoluble substances which may be present; and this distinction is often very important. We examine if the fusion of the bodies be effected slowly or readily, without any apparent motion, or with effervescence; if the glass resulting from the fusion be coloured, and if the colour be different with the oxidating flame from what it is with the reducing; lastly, we observe if the colour increase or diminish by cooling, and if at the same time the glass preserve or lose its transparency.

Certain bodies have the property of forming a clear glass with borax, which preserves its transparency after cooling, but, when slightly heated by the exterior flame of the lamp, becomes opaque and turns milk white, or is coloured, particularly if the flame has been directed on the glass in an unequal and intermitting manner. Such are alcaline earths, yttria, glucina, zircon, the oxides of cerium, columbium, titanium, &c. But one condition of this phenomenon is, that, to a certain point, the glass be saturated with the oxide or the earth. The same thing does not happen with silica, alumina, the oxides of iron, manganese, &c. and the presence of silica is sufficient to prevent the appearance of the phenomenon with those earths which form a glass with borax liable to become opaque, so that it does not ensue at all with their silicates, or only when the glass is supersaturated with them. To avoid prolixity, I shall say henceforth of a body presenting this phenomenon, that its glass becomes *opaque by flaming* (*au flamber*).

The use of borax is founded on the tendency of its component parts to form compounds that are all fusible, though in different degrees. On one hand it dissolves bases, and forms with them a fusible double salt with excess of base; on the other it dissolves acids, amongst which I place silica, and even to a certain extent alumina, and forms with them acid and fusible double salts. As all these salts commonly preserve their transparency on cooling, we can hence judge the more certainly of the colour which the compound acquires from the substance dissolved.

3. *Salt of phosphorus.*—To procure this salt, dissolve 16 parts of sal ammoniac in a very small quantity of boiling water, and add 100 parts of crystallized phosphate of soda; liquefy the whole together by heat, and filter the mixture whilst

boiling hot; the double salt will crystallize as it cools. The mother water that remains contains common salt, and a certain portion of the double phosphate. We cannot, however, obtain the latter by evaporating the mother water, for the common salt would crystallize with it; besides, the double salt becomes acid during the evaporation, and must, therefore, be saturated with ammonia when set to crystallize, if we would obtain a further portion from it. If the salt of phosphorus be not pure, it gives a glass which becomes opaque on cooling. In that case, it must be re-dissolved in a very small portion of boiling water and crystallized afresh.

The salt of phosphorus may be kept in large grains, or in powder. The crystals are generally of a convenient size for experiment. Heated before the blowpipe on charcoal, it boils up, intumesces a little, and gives out ammonia; acid phosphate of soda remains, which flows quietly, and forms a transparent colourless glass as it cools. It acts as a reagent principally by means of its free phosphoric acid, and the advantage of employing this salt, rather than the acid itself, is, that the latter is very deliquescent, much more costly, and is readily absorbed by the charcoal. The salt of phosphorus therefore shews the action of acids on the assays; its excess of acid seizes on all their bases, and forms with them more or less fusible double salts, whose transparency and colour may then be examined. Consequently, this flux is more particularly applicable to the examination of metallic oxides, whose characteristic colours it developes much better than borax.

The same flux exerts a repellent action on acids. Those which are volatile sublime, whilst the fixed acids remain in the mass, and divide the base with the phosphoric, or yield it entirely to it; in the latter event, they remain in suspension in the glass, without dissolving in it. In this respect the salt of phosphorus is a good reagent for the silicates; it sets the silica free, which then appears liquefied in the salt as a gelatinous mass.

4. *Saltpetre.*—Long thin crystals are to be selected, and kept in that state. The use of this reagent is very limited, and its object is to complete the oxidation of substances which have partly resisted the action of the exterior flame. This is effected by plunging the point of a crystal of saltpetre in the liquid mass; but in order to prevent the cooling of the globule, we have the crystal ready in a pair of pincers, which we hold between the third and fourth finger of the right hand (this does not prevent our managing the blowpipe with the same hand), and the moment we cease blowing, plunge it in the assay globule, and keep it there for an instant. The fused mass intumesces and becomes frothy; sometimes it assumes colour, whose tint becomes distinct on cooling; sometimes it exhibits no colour till completely cold. We need not continue the blast after the intumescence of the assay, for it is then that the effect is most distinct. Saltpetre is but little employed, except

to discover portions of manganese that are too minute to colour the glass without it; and since we have found still more delicate methods of accomplishing this, it is become almost useless.

5. *Vitrified boracic acid* is kept in coarse powder. Its use is limited, but necessary to detect phosphoric acid in minerals, by the process to be described when we treat of the action of the phosphates.

6 and 7. *Gypsum and fluor spar.*—These two substances, well dried, are used mutually to detect each other, and have no other use; but in this respect they are very interesting to the mineralogical chemist. If we put together a small morsel of gypsum with a rather smaller piece of fluor spar, and heat them, they soon melt at the point of contact, combine and are converted by fusion into a colourless transparent glass bead, which on cooling assumes the appearance of white enamel. In this way fluate of lime is used as a reagent for gypsum, and *vice versâ*. The compound resulting from the fusion of these two substances appears to be a double salt formed of fluoric acid, sulphuric acid, and lime, and since a somewhat larger volume of gypsum than of fluor spar is necessary, in order to obtain a very clear bead, it seems that it is composed of a particle of each of its constituent salts. If we take too much of either, the fusion is imperfect. If the glass bead be exposed to a strong and long continued heat, or be kept in fusion for some seconds before the

reducing flame, it fixes, swells up and is converted into an angular mass; after which it is no longer fusible. This phenomenon seems to result from the decomposition of the sulphuric acid, whence sulphurous acid is disengaged, and the double salt partially decomposed. Gypsum is not the only flux for fluor spar, nor fluor spar for gypsum; sulphate of baryta and fluor spar on the one hand, and fluate of baryta and gypsum on the other, also fuse very well together; but the fused mass is never clear, because the double calcareous salt which is formed, wants the property of dissolving the two barytic salts which it is mixed with.

8. *Nitrate of cobalt* dissolved in water. This solution must be very pure, entirely free from alkali, and rather concentrated.

This test is employed to detect the presence of alumina and magnesia, the former giving, after strong ignition, a fine blue with the oxide of cobalt, the second a pale rose colour. Silica does not prevent the appearance of these characters. There are two ways of treating a substance with nitrate of cobalt.

(*a.*) If the assay be absorbent, a small grain of it is used, which is to be moistened with a drop of the solution, and heated *strongly*, but *not fused*. After being heated some time, the assay becomes coloured, *blue* (more or less pure), if it contain alumina, and red or rose colour, if magnesia.

In the latter case, we must endeavour to fuse it,

for the magnesian compound retains its red colour after fusion, and generally even acquires a stronger tint. The blue colour of the alumina is also permanent in fusion, but it thereby loses its distinguishing character; for minerals which contain lime or alcali, without alumina, also become blue by *fusion* with oxide of cobalt; *but not till they have been fused.*

(*b.*) For harder substances, as the crystallized stones, the process is different. The stone is to be ground with water in the agate mortar, till reduced to the state of pap. A drop is to be taken up by the pestle, and laid on charcoal, which will absorb the water, whilst the fine powder suspended in it will remain on the surface. To this we add a drop of the solution of cobalt, and heat it gradually to redness, without stopping to examine the successive tints it passes through, before the salt is decomposed, and which end in black; for it is not till the moment of brightest incandescence, that the characteristic action is developed. If we perceive the mass begin to detach itself from the charcoal, we may take it up carefully with the platina forceps, and, exposing it in that state to the flame of the blowpipe, heat it more easily to the degree required.

The colour of the assay cannot be judged of till it is completely cold; and it must be examined in day light. The finest blue appears of a dirty violet colour, or almost red, by the light of a lamp or candle.

The proper quantity of the solution of cobalt to be taken depends on its concentration, and is easily learnt by experiment. The presence of a metallic oxide in the assay entirely destroys the action of this test, and if the solution contain any trace of saltpetre, it will impart a blue colour to certain substances, which otherwise would not exhibit it; for instance, silica and zircon.[1]

9. *Tin.*—This metal is employed in the state of foil, cut into long slips half an inch wide, and closely rolled up. Its use is to promote reduction in the highest degree in the fused vitreous compounds, especially when the assay contains small portions of metallic oxides, capable of being reduced to protoxides, and which in that state give more decisive results. We introduce into the still hot assay, pre-

---

[1] These phenomena had been observed by Gahn long before M. Thenard discovered the blue colour produced by alumina with arseniate or phosphate of cobalt; and they had led him to the discovery of the very same blue colour. He prepared it with alumina quite free from iron, obtained by precipitation from alum, over which he poured a concentrated solution of nitrate of cobalt, and exposed it, previously dried, to an intense heat. But, examining this beautiful colour by lamp light, Gahn found it rather red than blue, unpleasant to the eye, and did not consider it as fit to be used in the arts, nor even worthy of a particular description. This experiment was one of a series which he made to examine the modifications produced in minerals by heating them with metallic solutions: of them all, that of cobalt alone furnished an useful result. B. For Thenard's process for forming the alumino-cobalt blue, see his Traité de Chimie, vol. ii. p. 400. second edition. C.

viously exposed to the reducing flame, the extremity of the roll of tin, a part of which fuses and remains in the assay, and the whole is then immediately re-melted in the same flame. We must not continue the blast long, after having added the tin, for in that case it will either wholly precipitate the metal, which we wish to bring to the state of oxide, and all further action cease, or it will dissolve in so large a quantity in the flux (especially if it be salt of phosphorus), that the mass will become opaque.

10. *Iron*, used in the state of harpsichord wire, No. 6, 7, or 8. Bergman and Gahn employed it to precipitate copper, lead, nickel and antimony, and to separate them from sulphur, or fixed acids. For this purpose a small portion of one end of the wire is immersed into the fused assay, and the blast of the blowpipe directed on it, when the iron becomes covered with the reduced metal; the latter is often seen at the surface of the fused mass in the form of little globules. Iron has lately been applied to a more important purpose, founded on its property of reducing the phosphoric acid of the phosphates to the state of phosphorus, whence phosphuret of iron is produced, which, fusing with the assay, forms a white brittle metallic globule. When treating of the phosphates, I shall describe the process to be followed in this experiment.

11. *Lead* free from alloy, especially with silver, is employed in cupellation.

12. *Bone ashes.*—These are used with lead in cupelling metals or minerals containing gold or silver. The bone ashes must be reduced to a very fine powder, a small quantity of which is to be taken on the point of a knife moistened with the tongue, and kneaded in the left hand with a very little soda into a thick paste. A hole is then made in a piece of charcoal, and filled with the paste, and its surface smoothed by pressure with the agate pestle. It is then to be gently heated by the blowpipe, till it is perfectly dry. (The soda only assists the cohesion, and may be omitted.) The assay, previously fused with lead, is placed in the middle of the little cupel, and the whole heated by the exterior flame. When the operation is finished, the precious metals are left on the surface of the cupel. This experiment is so delicate that grains of silver visible to the naked eye, and indeed such as may be collected by the forceps, and extended under the hammer, may in this way be extracted from the lead met with in commerce.

I have seen tobacco pipes used for this purpose in England, on a transverse section of which the operation is performed. They have the disadvantage of absorbing very little oxide of lead, and of allowing us to work only on very minute portions of matter, which proportionately diminishes the button of silver. I, for a time, tried the use of burnt round bones, performing the cupellation on one end, and carefully removing the portion loaded with oxide of lead after every operation. But from the extreme brittleness of burnt bones, the process first described is by much the best.

13. *Silica,* such as we obtain in the analysis of minerals, and particularly from pulverised rock crystal, forms a very fusible glass with soda, by means of which we may detect the presence of sulphur, or sulphuric acid. I sometimes use glass instead of silica, but it does not mix so readily with the soda. For the rest, see further on, the article *Sulphates.*

14. *Oxide of copper* is used to detect the presence of muriatic acid. See the article *Muriates.*

15. *Test papers,* respectively tinged with infusions of litmus, brazil wood, and turmeric.

General rules for experiments with the blowpipe.—It sometimes happens that an assay with which we have been a good while occupied, is suddenly blown off from the support. To recover it, Gahn placed his lamp in a large tray made of sheet iron, not tinned, its border about an inch high, and its bottom covered with a sheet of thick white paper. A white earthenware dish, or a paper tray, will be a good substitute for the iron one in travelling. The tray should be emptied after every experiment, to avoid confounding the assay with other matters.

As to the size of the morsel operated on, it is large enough, if we can distinctly see the effects produced on it, and we are more likely to fail in our object by using too large, rather than too small a piece.

Von Engeström directs the assay piece to be about the size of a cube 1-8th of an inch on the side, which would be very well if we were working with an enameller's lamp, but is much too large for our experiments. Bergman says, about the size of a pepper corn, adding, that we must often operate on smaller portions. This also is many times too large; we should seldom succeed in applying an adequate heat to so considerable a bulk, and even for vitrification with the fluxes, the size of a pepper corn is much too great. A piece of the size of a large grain of mustard seed, is almost always sufficient. For the rest, experiment must teach us to determine the bulk best adapted to the success of an experiment, particularly when we attempt to produce an effect on a small quantity which has failed with a larger. But, even if the experiment succeed with a large piece, it always requires a stronger and a longer blast, and after all, we can judge of the colour and fusibility just as well with a smaller. The only instance in which it may be convenient to operate on portions larger than a mustard seed, is when we wish to extract metals, whether by reduction with soda or by cupellation, because in that case we obtain a larger portion of the metal sought for, which may consequently be examined and distinguished with greater ease.

Prior to submitting an assay to the action of the

fluxes, the phenomena it presents, when heated alone before the blowpipe, must be examined as follows.

(*a.*) The substance must be heated in a small mattrass by the spirit lamp, to ascertain if it decrepitate, or give off water or any other volatile matter.

(*b.*) We must heat the assay gently on charcoal by the flame of the oil lamp directed on it by the blowpipe, and immediately after removing it from the heat, apply it to the nose to ascertain if it contain any volatile acid, arsenic, selenium, or sulphur. The odour developed by the oxidating flame must be compared with that which the reducing flame developes, and the difference noted down. The first renders the odour of selenium and sulphur more sensible, the second that of arsenic.

(*c.*) The fusibility of the assay must be examined. If we can obtain it only in the form of round grains, the best way is to place them on charcoal, notwithstanding their liability to be blown off when not very fusible. But if we can choose a form to our liking, we strike off from the specimen by the hammer, a very thin scale, one edge of which is usually transparent; or select amongst the detached fragments, a pointed or lamellar morsel, which we hold in the platina forceps and present its edge or point to the flame. We thus immediately perceive if the assay be fusible or not. Infusible substances retain all the sharpness of their points and edges, which is easily distinguished by the microscope. With difficultly fusible substances, the same parts become rounded; and, lastly, those of easy fusion are melted into a globule. As to very difficultly fusible minerals, I generally grind them with water, and put a drop of the mixture on charcoal, precisely as is done in the experiments with nitrate of cobalt; I then dry the mass spread over the surface of the charcoal, and expose it to the oxidating flame, till it adheres no longer to the support. It then forms one coherent body, a sort of cake, which I take between the platina forceps, and give its edges the greatest heat I can produce. The edges, even of substances that I call infusible, usually become slightly curved, which is a proof that, strictly speaking, they are not absolutely infusible, but the microscope shows us whether they be vitrified or not. In the same manner we treat dry pulverulent substances, after having made them into a paste which is spread on the surface of the charcoal with the point of the knife.

I am convinced that the heat we can produce by the blowpipe supplied with air from the lungs, is very limited; that we cannot, for instance, fuse by its means the smallest portions of either alumina or silica, and, consequently, that the determination of differences of fusibility does not depend so much on the size of the assay and the skill of the operator, as has been supposed. In this respect the blowpipe has a decided advantage over the machines for producing a current of oxygen gas.

H. de Saussure made a series of experiments to determine the relative fusibilities of mineral substances, and calculated in degrees of Wedgewood's pyrometer, the temperatures at which these different bodies begin to fuse, from the proportion which the diameter of the largest volume of each mineral that he could fuze into a globule, bore to the diameter of the largest globule of silver fusible at a known temperature. His tables are certainly very valuable, but since they cannot furnish useful indications in such experiments as we are treating of, and as the results they present are at most but approximations, I shall say no more about them.

Certain substances, particularly certain minerals, undergo a change of form and aspect before the blowpipe, without, however, entering into fusion. Some swell up like borax, or form ramifications which, taken altogether, have the appearance of a cauliflower. Of these substances, some fuse after intumescence, others retain that state without fusing. Other mineral substances throw out a sort of foam whilst fusing, and afford a *blebby* glass, which, although very transparent in itself, produces in mass the effect of dull glass, from the multitude of air bubbles which it contains.

This intumescence and foaming do not take place, in most minerals, till they have acquired the temperature at which the whole of their water is driven off: the ramifications seem to result from a new state of equilibrium between the constituent parts of the bodies, produced by the heat; as to the intumescence, and the production of foam which ensues after fusion, they can only belong to the expansion of a component part of the substance which is volatile and assumes the gaseous form, although these phenomena often occur with compounds in which analysis fails to detect the presence of any similar matter. They take place chiefly in the double silicates of lime or alcali, and alumina. Sometimes they disappear after a few moments' blowing; at others they last as long as the substance remains liquid. In the latter case it appears that the assay takes carbonic acid from the flame, which is transformed by the contact of the charcoal into carbonic oxide, and that it is this which fills the blebs. The cause of these various kinds of intumescence deserves particular investigation; as long as we remain ignorant of it, we cannot flatter ourselves with having a thorough knowledge of the bodies that present this species of phenomenon. In the mean time it affords a good character by which to distinguish substances that in other respects are alike.

In the use of fluxes, we must be careful not to suspend the blast too soon. A substance may seem to be infusible at the beginning, which by degrees yields to the action of the flux, and at the end of two minutes enters into perfect fusion. Moreover, we must put only small portions of it to be fused at a time, and wait till the first has undergone the action of the flux, before we add a second, so that, in short, the glass produced may attain a degree of saturation beyond which it can dissolve no more of the assay; when the glass is thus saturated it often presents manifest effects which could not be produced with the glass not saturated.

When operating with fluxes before the reducing flame, it sometimes happens that the assay globule oxidates as the charcoal cools, and we thus lose the fruit of our first process. To obviate this inconvenience, we turn the charcoal upside down, so that the globule, still fluid, may fall on some cold substance, as the iron tray under the lamp, or, if that cannot be done, we pour a drop of oil over it. The use of the oil, however, is productive of another inconvenience,—it frequently carbonizes, and thus renders the glass obscure, which should be avoided.

If the colour of the fused assay be so intense as to appear opaque, we may ascertain its transparency by holding the globule in a certain direction opposite the flame of the lamp, when we may distinguish, even in broad day-light, the reversed image of the flame painted on the glass, in the colour of the latter. We may also flatten the glass, before it becomes solid, between a pair of forceps with flat claws, previously heated, and if that be not sufficient, we must endeavour to draw out the glass globule, at the instant it begins to cool, into a thread fine enough for its colour to be seen by transmitted light.

According as they are exposed to the outer or inner flame of the lamp, and fused alone or with fluxes, mineral substances present numerous phenomena which must be carefully noted, and form, when taken together, the general *result* of the trial to which each individual has been submitted. The minutest circumstance of these phenomena must be attentively observed, because it may often lead to the detection of elements whose presence was not suspected.

With regard to quite new substances, the advantage which the operator may derive from the blowpipe in ascertaining their presence and some of their properties, depends entirely on the extent of his knowledge of chemistry in general, and pyrognostic phenomena in particular, as well as on his personal skill in observing and seizing on whatever is characteristic in the effects produced. On this subject no particular rule can be laid down.—Whenever we would record the result of an experiment with the blowpipe, either for our own instruction or that of others, we must always make two experiments, note down separately the result of each, and then compare the two together, for it often happens that something which escaped us on a first observation, strikes us on a second. The safest mode is for two persons to make, and note down separately, a similar set of experiments, and compare their results; if they agree, they may be considered as accurate, otherwise the cause of discrepancy must be sought for. A little difficulty sometimes attends this sort of association, from two persons not always seeing and denominating colours alike. For instance, there were certain shades which Gahn always called yellow, or dull yellow, and which I persisted in calling red, although we agreed as to their fundamental colours, pure yellow and pure red.

DESCRIPTION OF THE PHENOMENA PRESENTED BY DIFFERENT MINERAL SUBSTANCES BEFORE THE BLOWPIPE.

We have nothing to do in this place with the non-metallic simple substances, their physical characters are sufficient to distinguish them in the elementary state, and their properties are supposed to be known to those who seek them in their compounds. Of these compounds, those with which we shall first employ ourselves, are the combinations of their oxides and acids with oxidated bodies; they will be described under the article *Salts*.-

I shall begin, therefore, with the metals, treating principally of their oxides (amongst which I class the alcalies and earths), and detail what I know of the methods by which they may be distinguished. With this view I have made a series of new and particular researches, in which each substance has been treated individually. In the course of my experiments I have discovered, in respect to many of them, characteristic phenomena which I long sought for in vain, and could never have obtained without that continued attention which my task, taken as a whole, required. I am persuaded that when the use of the blowpipe shall become more general, others still more precise will be discovered. I shall now describe particularly the phenomena produced by the different oxides in their pure state, or, which is pretty much the same as to experiments with the blowpipe, in the state of carbonates or hydrates.

A. OF ALCALIES, EARTHS, AND METALLIC OXIDES.

1. *Alcalies.*

The alcalies do not afford indications, when acted on by the blowpipe, by which they may be ascertained with perfect certainty. When pure or in the state of carbonate, their taste, or their action on litmus paper reddened by an acid, is still the best test. When in combination in salts, their presence is ascertained by fusing the salt with soda on platina foil; if no precipitation ensue in the liquified mass, we may conclude that the base of the salt is an alcali. But if the same compound contain both an earth and an alcali, I know no method of ascertaining the presence of the alcaline part by the blowpipe. Neither does it furnish us with any characteristic sign by which to distinguish *potassa* from *soda,* unless oxide of cobalt be present. (See further on, *Oxide of Cobalt.*)

Lithia differs from the two alcalies of which I have just spoken, in this, that when fused on platina foil, it attacks that metal, and leaves round the spot on which it lay a dull yellow trace. The salts of lithia do not produce the same effect unless soda be added, which, by disengaging the lithia, leaves it free to act on the platina; but this is a very equivocal character, for although it takes place in a decided manner with a salt of lithia, other bodies also which contain no lithia produce it to a certain extent when heated with soda; and soda itself sometimes leaves a trace on platina round the spot where it lay. The same phenomenon does not occur with potassa; on the contrary, when added in excess to the assay, it prevents the action of the lithia contained in it. The yellow spot may be removed by washing the platina with water and heating the metal red hot, but we perceive afterwards by the frosted appearance of the part, that the metal has lost its polish on that spot; this difference of aspect is particularly sensible during ignition.

Ammonia is seldom an object of experiment with the blowpipe. In general we need only mix a little soda with the substance supposed to contain ammonia (which indeed may then be immediately perceived by the odour of the mixture), and heat the whole gently in a matrass; a sublimate of carbonate of ammonia will be formed. We may also heat the assay separately in a glass tube closed at one end, and introduce a piece of moistened and slightly reddened litmus paper.[1]

2. *Baryta.*

Alone does not fuse, till it has absorbed a little

---

[1] In default of sufficient characters to distinguish *potassa*, *soda* and *lithia* by the blowpipe, it seems desirable to state the tests by which each may be known in solution. *Potassa* and its salts give with muriate of platina a fine crystalline yellow precipitate inclining to orange. The same test produces no effect with solutions of soda or lithia, or their salts.

Carbonate of lithia is much more insoluble than either carbonate of potassa or soda, requiring at least 100 parts of water for its solution. If this be converted into a muriate and evaporated to dryness, the residual mass is soluble in alcohol, and imparts a crimson colour of great beauty to its flame. The negative property of soda with regard to muriate of platina and alcohol, will serve to distinguish it from both the other fixed alcalies. I may add that the alcaline salts should be heated sufficiently to expel all the ammoniacal salts, if any be present, before we add the muriate of platina, as the latter have a similar action on that test to the salts of potassa. C.

water produced by the flame.² Its *hydrate* fuses, bubbles up, intumesces and ends by solidifying at the surface, after which it penetrates with a brisk bubbling into the mass of the charcoal, where it soon loses its water and is transformed into a solid crust.

Carbonate of baryta fuses very readily into a clear glass, which, on cooling, assumes the aspect of a white enamel; on charcoal, it ends by effervescing strongly, sputters up, becomes caustic and is absorbed with the same phenomena as the hydrate. Both the hydrate and carbonate, as well as pure baryta, produce the following effects with the fluxes.

With borax, baryta fuses readily with brisk effervescence into a clear glass. If we increase the quantity of baryta we obtain a clear glass, which on cooling becomes covered towards its base with little milk-white tubercles; a still further quantity gives a clear glass which turns first milk-white, and lastly enamel white, beginning at

---

[2] I shall subjoin in this, and several other instances, the appearance of refractory substances, after fusion by the gas blowpipe, described p. 16. I give the facts as I find them in Dr. Clarke's account of the "Gas Blowpipe;" London, 1819.

Baryta.—" It became fused very readily, and assumed the form of a *jet black shining slag ;* its [fusion being accompanied with a *chrysolite green coloured flame,* and in some instances with a slight degree of scintillation; at the same time dense *white fumes* were evolved, and the supporter" (forceps made of slate) "became invested with a white oxide." Gas Blowpipe, p. 69. C.

the base. The glass, which continues clear on cooling, becomes opaque by *flaming*.

With salt of phosphorus baryta fuses readily with brick effervescence, during which the globule foams and swells up; the intumescence then subsides, and the assay becomes a clear glass. By increasing the dose of baryta we obtain a clear glass, which on cooling is covered here and there with milk-white spots; by a further addition the glass assumes the appearance of enamel on cooling.

With nitrate of cobalt baryta gives a globule which, whilst hot, has a red brown, brick red, or rusty yellow colour, according to the quantity of nitrate; the colour disappears on cooling. When heated again the colour does not return till the assay is in fusion; exposed to the air, it soon falls into a clear grey powder.

3. *Strontita.*

Alone, in the state of hydrate, it presents the same phenomena as baryta.[1] Carbonate of strontita

---

[1] " This oxide being much more refractory than the preceding" (baryta), "is almost *infusible per se* even with the aid of the gas blowpipe." Dr. Clarke, therefore, mixed it into a paste with oil, &c. and after repeated exposure to the " gaseous flame," it began to " exhibit the appearance which the barytes assumed after its fusion, namely, a jet black shining substance." Dr. Clarke adds, that during the whole of the experiment, the flame is tinged with an intense amethystine purple colour. The fusion is attended with scintillation and the evolution of acrid and suffocating dense

fuses with a moderate heat at the surface, and begins to assume a ramified cauliflower appearance of dazzling brightness; in an intense reducing flame it imparts a feeble tint to it, not easily perceived by day light; the ramified portion has the taste and effect of an alcali.

With borax and salt of phosphorus it behaves like baryta.

Soda does not dissolve caustic strontita; carbonate of strontita, mixed with an equal volume of soda, fuses into a clear glass which becomes milky white on cooling; in a stronger heat the mass bubbles up, becomes caustic and sinks into the charcoal. A larger quantity of the carbonate of strontita does not wholly fuse, but becomes caustic with a strong heat, and is absorbed by the charcoal.

With solution of cobalt strontita exhibits a black or greyish black colour, and does not fuse like baryta.

4. *Lime.*

Alone, caustic lime neither fuses nor suffers any change;[1] carbonate of lime readily becomes caustic,

---

white fumes, which the operator must be careful not to inhale." Gas Blowpipe, p. 73. C.

[1] "Lime in a state of perfect purity, and in the pulverulent form, being placed within a platinum crucible, and exposed to the flame of the gas blowpipe, its upper surface became covered with a *limpid botryoidal glass*, resembling

and throws off a brilliant white light; it acquires alcaline properties, and falls to powder on being moistened.

With borax fuses readily into a transparent glass, which becomes opaque by *flaming*. The carbonate fuses with effervescence; a larger quantity of lime gives a transparent glass, which crystallizes on cooling, but not in so regular facets as phosphate of lead. The glass never assumes so perfect a milky whiteness, as those of baryta and strontita.

With salt of phosphorus lime fuses in large quantity into a clear glass, which remains transparent on cooling; the carbonate fuses with effervescence and gives the same result; if we add a fresh portion of carbonate of lime to the mass, the carbonic acid flies off and the new portion is converted into phosphate of lime without losing its form. By a long continued blast, the phosphate of lime itself dissolves and, on cooling, deposits small crystalline needles on the part not fused; if the glass be completely saturated it becomes perfectly milk-white on cooling.

Soda dissolves no sensible portion either of lime or its carbonate, but sinks into the charcoal, leaving a slightly rounded calcareous mass on the surface.

---

hyalite; the inferior surface was quite *black*. Its fusion was accompanied by a lambent *purple* flame." Gas Blowpipe, p. 47.

Iceland spar fused, with great difficulty, by the same means, into a "brilliant limpid glass." Ibid. p. 48. C.

With *solution of cobalt* lime gives a black or dark gray infusible mass.

5. *Magnesia.*

Alone, undergoes no alteration.[1]

With borax behaves like lime, but does not crystallize so perfectly.

With salt of phosphorus fuses readily into a clear glass, which, when saturated with magnesia, becomes milk white on cooling. If not quite saturated it takes that colour by *flaming*.

With soda, no action.

With solution of cobalt, after a strong blast, assumes a fine flesh colour, the tint of which is very feeble and not easily distinguished till the assay is perfectly cold.

6. *Alumina.*

Alone, appearance not altered.[2]

[1] *Pure magnesia*, "fusion *per se* extremely difficult. When the powder is made *to* adhere (by moisture with distilled water and subsequent desiccation) and placed upon *charcoal*, it is fusible into a *whitish glass*, but the parts in contact with the *charcoal* acquire an imposing *pseudo-metallic* lustre."

Hydrate of magnesia. (*Pure foliated magnesia* from America.) "This substance is incomparably refractory; with the utmost intensity of the *gas blowpipe*, it is ultimately reducible to a *white opaque enamel*, invested with a thin superficies of *limpid glass.* Its fusion is accompanied with a purple coloured flame." Gas Blowpipe, p. 53. C.

[2] *Pura alumina.* "Fusible without difficulty into a *snow white opake glass.*" Gas Blowpipe, p. 54. C.

With borax fuses slowly into a transparent glass, which does not become opaque, either by cooling or *flaming;* if we add to it a large portion of alumina in fine powder, we obtain an opaque glass, whose surface is crystalline on cooling, and which becomes almost wholly infusible. This glass is opaque both cold and hot.

With salt of phosphorus alumina also fuses into a transparent glass, which does not become opaque at any degree of saturation; if the proportion of alumina be too large the unfused portion is semi-transparent. A mixture of salt of phosphorus and borax, both saturated with alumina, gives a clear glass.[1]

With soda swells up a little, forms an infusible compound, and the excess of soda sinks into the charcoal.

With solution of cobalt, with a strong blast, gives a fine blue colour, which becomes deeper, without losing its beauty, by an additional quantity of cobalt; it is not distinctly seen but by day light, nor until the assay be cold.

7. *Glucina.*

Alone, no change.

Borax and *salt of phosphorus* dissolve a large proportion of glucina, and convert it into a clear

---

[1] Those glasses which turn milk white by *flaming,* remain opaque when we mix that formed with salt of phosphorus with the one formed with borax. B.

glass, which becomes milk white by *flaming;* if we add a further quantity of glucina, the glass becomes milk white on cooling.

With soda, no action.

With solution of cobalt glucina forms a black or dark grey mass.

8. *Yttria.*

Its characters resemble those of glucina.

9. *Zirconia.*

Alone, zirconia, as obtained by calcining its sulphate, emits a more brilliant light than any other substance I have tried; the whole assay is illuminated by it, and its brilliancy is so dazzling, even in broad day light, that the eye can scarcely support it. It is absolutely infusible. Klaproth says that zirconia agglutinates before the blowpipe, in consequence of incipient fusion; in that case it is not pure.

With borax, salt of phosphorus, and *soda,* its habits resemble those of glucina, except that it dissolves with greater difficulty in salt of phosphorus, and more readily forms an opaque glass with that re-agent.

10. *Silica.*

Alone, no change.[1]

---

[1] H. de Saussure states, that he fused silica (rock crystal) by the flame of a large candle, blown with a pair of double

With borax fuses very slowly into a clear glass, difficulty fusible and incapable of being made opaque by flaming.

Salt of phosphorus dissolves but a very small quantity of silica; the glass retains its transparency when cold; the unfused portion is semi-transparent.

With soda fuses with brisk effervescence into a clear glass.

With solution of nitrate of cobalt, in a certain proportion assumes a feeble bluish tint, which changes to black or dark grey, by an additional quantity of cobalt. By this colour silica may be distinguished from aluminous substances. If we expose a thin portion of the assay to a strong heat,

bellows, the area of whose surface was 62 square inches; the silica was supported on *disthene*, and converted into a globule of 0.014 diameter. I have never been able to fuse the thinnest lamina of silica, either on charcoal or in the forceps; and I suspect that in Saussure's experiment the action of the support on the assay on the one hand, and the greater purity of the air which supplied the flame on the other, contributed to a result that could never be obtained by the common blowpipe. B.

Rock crystal.—" The most highly diaphanous specimen that could be procured was exposed to the flame of the *gas blowpipe* with perfect success. In the first trial, the edges only were fused, and resembled hyalite. In the second trial, the fusion was completed; the crystal then appeared in the form of one of *Prince Rupert's drops*, having lost nothing of its transparency, but being full of bubbles." Gas Blowpipe, p. 63. C.

that portion fuses with the oxide of cobalt into a blue glass; the part next the fused portion also becomes blue, but this colour approaches to red, and has nothing pleasant to the eye.

11. *Molybdic Acid.*

Alone, in an inclined tube, it fuses and gives off fumes which condense partly on the sides of the tube, covering it with a white powder, and partly on the surface of the fused mass in brilliant pale yellow crystals. Heated on platina foil, this acid fuses and gives off fumes; the fused part has a brown colour, but becomes yellowish and crystalline on cooling; in the reducing flame it becomes blue; a violent heat turns it brown. On charcoal it fuses, and is absorbed; a portion of the acid revives by continuing the blast, and the metal may be collected, by pulverising and washing the charcoal, in the form a grey metallic powder.

With borax, on the platina wire, it fuses in the exterior flame into a colourless transparent glass. Before the reducing flame on charcoal, the glass becomes dirty brown without losing its transparency, and not unlike that which we obtain by a mixture of the peroxide and protoxide of iron; if we add a fresh portion of molybdic acid to the mass, the glass becomes opaque in the reducing flame, and a multitude of brown scales of oxide of molybdena may be perceived in its interior, apparently surrounded by a slightly brownish transparent glass.

With salt of phosphorus it fuses before the exterior flame on the platina wire into a transparent glass, inclining to green whilst hot, but which loses its colour on cooling; in the reducing flame the glass becomes opaque and appears black or dark blue, but on cooling it becomes limpid and shews a fine green colour, which almost rivals that of the oxide of chrome. The same colouring effect takes place on charcoal (particularly if the proportion of acid be considerable), both in the exterior and interior flame. On the platina wire, the green glass may be oxidated by the exterior flame, when it recovers its transparency on fusing. The reducing flame is incapable of separating any portion of the brown oxide of molybdena from the salt of phosphorus. Tin does not change the colour of the reduced green glass, but the metal is observed to swell up in the glass and, to a certain extent, combine with the molybdena.

With soda, on the platina wire, molybdic acid effervesces and fuses into a clear glass, which becomes milk white on cooling; if the acid *be in small quantity*, before the reducing flame the glass assumes nearly the same colour as glass of borax in similar circumstances, but becomes opaque on cooling; if, on the contrary, the glass be *supersaturated with molybdic acid* and exposed to a strong reducing flame, the acid is partly reduced to the state of oxide, partly to the metallic state; and if we dissolve the saline mass in water, a greyish brown residium of considerable specific gravity is obtained, which under the steel polisher exhibits metallic brilliancy and the grey colour of iron.

Treated with *soda*, on charcoal, molybdic acid is absorbed as soon as fused, and a large quantity of molybdena, in the state of a steel grey metallic powder, may be collected by pulverisation and washing; if we place on the spot where the assay is absorbed, a fresh portion of molybdic acid with a very small quantity of soda, and expose it to a good reducing heat, the new assay remains on the surface of the support, and we obtain a grain of reduced molybdena and molybdate of soda, from which the pure metal may be separated by washing. Molybdena, therefore, contrary to the general opinion, ranks with the easily reducible metals, although it does not undergo fusion at the same time that it is reduced.

12. *Tungstic Acid.*

Alone, it blackens, but is not fused by the reducing flame.

With borax it fuses readily on the platina wire, before the exterior flame, into a transparent colourless glass which cannot be made opaque by flaming. With a small proportion of acid in the reducing flame the glass becomes yellowish, the colour increasing in intensity as it cools, till it is quite yellow. If we add a fresh quantity of acid at the instant it congeals and is converted into a frothy mass, in consequence of the liberation of gas; the intumescence disappears on a second fusion, but the mass swells up afresh on cooling. This phenomenon ensues whether the assay be acted on by the exterior or interior flame, and as well on charcoal as on the platina wire; I have not been able to discover the cause of it: it does not take place when the glass is transparent.

Soda gives a dark orange coloured glass with oxide of chrome before the exterior flame on the platina wire, which becomes opaque and yellow on cooling. In the reducing flame the glass becomes opaque; after cooling it is green. Charcoal absorbs it, but I have never been able to discover any trace of reduced metal in the absorbed mass.

14. *Antimony and its Oxides.*

Antimony in the metallic state fuses readily on charcoal; when heated red, it remains ignited a considerable time without the action of the blowpipe, giving off dense white fumes. The fumes are gradually deposited and form round the metallic globule a sort of network of little crystals, having a pearly lustre, which at length cover it completely. For a few seconds we may perceive, by the light of the lamp, the metal in the interior of the mass in a state of ignition, after it is perfectly enveloped by the crystalline network of oxide of antimony. This oxide fuses when the flame of the lamp is directed on it. Metallic antimony, heated alone in a matrass, does not sublime at the temperature at which glass melts. Heated to redness in an open tube, it burns slowly, sending out white fumes which condense on the glass, and here and there present traces of crystallization. These fumes are nothing more than the oxide of the metal, and may be driven by heat from one part of the tube to another, without leaving the least residuum; but if the antimony be combined with sulphur, it forms, besides the oxide, a certain portion of antimonious acid, which adheres to the surface of the glass in the form of a white coating, after the oxide has been driven off by heat.

Oxide of antimony alone fuses easily and sublimes in the form of a white vapour. This oxide, such as is obtained by precipitation, washing and drying, often inflames before it fuses, burns like tinder, becomes infusible and is converted into antimonious acid. On charcoal it is reduced to the metallic state, colouring the flame greenish in the process.

Antimonious acid does not fuse, but gives out a bright light and diminishes in bulk in the interior flame, whilst the charcoal is covered with white fumes: but it is not reduced like the oxide.

Antimonic acid whitens on the first impulse of the heat and is converted into antimonious acid; if it contain water it gives it off and passes from white to yellow; it then by ignition gives off oxygen and resumes its white colour.

The oxide and the acids of antimony present the same phenomena with the fluxes.

Borax dissolves a large quantity of antimonious acid without becoming opaque. The glass has a yellowish tinge whilst hot, but loses it in great measure on cooling. When saturated, some metallic antimony sublimes and condenses on the charcoal round the assay; if the glass be strongly heated by the reducing flame it becomes opaque and greyish, in consequence of the metallic particles formed in it.

Salt of phosphorus forms a transparent colourless glass with the acid. On the platina wire, in the oxidating flame, the glass acquires a feeble yellowish colour which disappears on cooling. If it contain iron, it assumes in the reducing flame the same red colour as tungstic acid, and the ferruginous oxide of titanium. By a strong blast the antimony is reduced and evaporates, and the colour disappears. The addition of tin produces a similar effect.

With soda it fuses on the platina wire into a transparent colourless glass, which turns white on cooling. It is reducible on charcoal.

15. *Oxide of Tellurium.*

Alone, on platina foil, it fuses and gives off fumes; on charcoal it fuses and is reduced with effervescence. The reduced metal may easily be confounded with bismuth and antimony; under the article bismuth, I shall shew how they are distinguished.

With borax and the salt of phosphorus on the platina wire, it gives a clear colourless glass, which becomes grey and opaque on charcoal, from a great number of particles of reduced metal.

With soda on the platina wire, it gives a colourless glass, which becomes white on cooling. It is reducible on charcoal.[1]

16. *Oxide of Columbium.*

Alone, no change.

With borax it forms a colourless transparent glass, which becomes opaque by *flaming*, and, if the proportion of oxide be large, enamel white on cooling.

With the salt of phosphorus it fuses readily, and in large quantity into a colourless glass, which retains its transparency on cooling

With soda, it combines with effervescence, but without the solution or reduction of the oxide.

Since the oxide of columbium in these respects resembles the earths, properly so called, it might easily be confounded with some of them, in an experiment with the blowpipe. Its combination,

---

[1] It is not easy to collect the little globules of tellurium into one. In larger operations, the addition of a portion of nitre thrown on the hot metal in the crucible or retort, will assist its fusion into a mass. C.

however, with the salt of phosphorus does not becomes opaque on cooling, even though the oxide of columbium be in excess; whereas the contrary is the case with glucina, yttria and zirconia. If we add a fresh excess of the oxide to the salt of phosphorus, it spreads itself uniformly over the surface of the glass, which then becomes opaque, even whilst in fusion; but the part not fused does not turn milk white; it becomes semi-transparent like silica, from which however it differs in its action with soda. Oxide of columbium is distinguished from alumina by the phenomena it produces with borax, and the solution of cobalt; the latter gives no blue colour with oxide of columbium.

17. *Oxide of Titanium.*

Alone, it undergoes no change.

With borax on the platina wire, it fuses readily into a colourless glass, which becomes white milk by *flaming*. If the proportion of the oxide be increased, the glass turns white spontaneously on cooling. If the quantity of oxide be small, the glass first becomes yellow in the reducing flame, and when the reduction is complete it assumes a dull amethyst colour, which becomes more distinct on cooling. This glass is transparent and a good deal resembles that of the oxide of manganese acted on by the oxidating flame, but inclines rather more to blue. With a larger proportion of the oxide, the glass becomes dull yellow on charcoal in the reducing flame, and on cooling acquires so deep a blue colour, that it appears black and opaque. If it be then *flamed*, it becomes light blue, but opaque and like enamel. The beauty of the blue tint varies in different assays. These phenomena are occasioned by the glass containing both peroxide and protoxide; the latter tends to give it a dark blue colour, and the first, which does not contribute to its colour, causes it to assume on *flaming* the appearance of a white enamel; from the mixture of the white and the deep blue, arises the light blue, whose intensity depends entirely on the relative quantities of the peroxide and protoxide in the glass; if that of the former be small, the glass remains black; but if it contain only a little of the protoxide, it whitens by *flaming*.

Salt of phosphorus dissolves oxide of titanium in the exterior flame, and converts it into a clear colourless glass. In the reducing flame, this oxide gives a glass which appears yellowish whilst hot, but on cooling, at first reddens, and at last assumes a very beautiful bluish violet colour. With too large a quantity of oxide the colour is so deep that the glass seems opaque, without, however, having the appearance of enamel. The colour may be discharged by the exterior flame. The reduction is more easily effected on charcoal than on the platina wire, but even on charcoal a well sustained flame is necessary, particularly in the assay of *minerals* containing titanium, as sphene. The addition of tin considerably facilitates and accelerates the reduction. If the oxide of titanium contain iron, or if we add iron to a glass coloured by oxide of titanium, the violet colour derived from the protoxide disappears, and in the reducing flame, the glass assumes a red colour similar to that developed by the ferruginous tungstic acid. If these substances be in small quantity the colour becomes yellowish red, but it does not appear till the glass begins to cool, and generally does not acquire its full intensity till the globule is perfectly cold. Such is the delicacy of this test, that when the glass contains so little oxide of titanium that we cannot decidedly ascertain its presence by examining the colour, we may immediately perceive it by adding iron, particularly metallic iron, when the effect is infallibly and unequivocally produced. The glasses of ferruginous tungstic acid, ferruginous antimonious acid, and of oxide of nickel, also assume the same shade in the reducing flame; but it is easy to ascertain which of these substances is combined with the iron. A few moments' good blast with the reducing flame is sufficient to drive off the antimonious acid; which done, only the tint derived from the iron remains. I have already said, that the ferruginous tungstic acid gives a glass with tin sometimes green, sometimes blue; now if the glass of ferruginous titanium be treated with tin, the colour derived from the iron disappears, and the violet tint of the protoxide of titanium re-appears. That this effect, however, may be produced, it is essential that the colour of the glass be not very intense; if it be, a fresh portion of the flux must be added. But, since it often happens that the colour in question is developed by so small quantity of the ferruginous oxide of titanium, that the oxide of titanium it contains could not alone have produced any sensible effect, in such case, tin entirely destroys the colour and the action is annihilated. We must then form a glass with the assay and the salt of phosphorus, which must be completely saturated with the substance, and treat it with tin; in this way we may frequently succeed in developing the colour derived from the oxide of titanium, particularly after the glass is quite cold. The effect produced by the oxide of nickel is distinguished from the preceding, by its attaining its maximum when the glass is hot, and disappearing almost wholly on cooling, and by its being the same both in the exterior and interior flame, whereas the effect produced by the other substances disappears in a good oxidating flame. The modification produced by iron on oxide of titanium and tungstic acid, does not take place with borax.

With soda, oxide of titanium fuses with effervescence and sputtering, into a dull yellow transparent glass, which is not absorbed by the charcoal, and becomes white, or greyish white on cooling. This glass has the property of crystallizing at the moment it ceases to be ignited, disengaging at the same time so much heat, that the globule ignites afresh, and even becomes white hot. The same phenomenon occurs with all bodies that crystallize at a high temperature, as, for example, phosphate of lead, but I have never seen so much heat developed by crystallization, as in the preceding instance. The intensity of the phenomenon depends chiefly on the soda and oxide being exactly in the proper proportions. If we employ more oxide of titanium than the soda can dissolve, so that the parts not fused remain suspended in the glass, the phenomenon does not take place; but if we then add small successive portions of soda, till the exact quantity necessary to fuse the oxide be attained, it appears in all its brilliancy. A further quantity diminishes the effect, and with a large excess of soda the whole mass is absorbed by the charcoal.

Oxide of titanium is not reducible with soda on charcoal. I used in my experiments the oxide of titanium extracted from French rutilite by Laugier's process, and in every trial to reduce it I obtained some flattened grains of a white, unmagnetic malleable metal, having all the appearances of tin; I found afterwards that when this oxide of titanium has been digested in hydrosulphuret of ammonia, that reagent separates oxide of tin from it by solution, which remains after the water has been evaporated and the dried mass roasted, and is easily reduced to a metallic globule.

With solution of cobalt, the oxide of titanium assumes a black, or greyish black colour.

18. *Oxides of Uranium.*

Alone, the peroxide of uranium blackens, and passes to the state of protoxide, but does not fuse.

With borax, it fuses into a dark yellow glass, which becomes dirty green in the reducing flame. The yellow colour may be restored by the oxidating flame on the platina wire; on charcoal the operation is very difficult. The green glass, at a certain degree of saturation, becomes black by *flaming*, but neither crystalline, nor like enamel.

With salt of phosphorus on the platina wire, it fuses in the oxidating flame into a transparent yellow glass, whose colour fades on cooling, and at last becomes straw yellow with a slight tinge of green. In the reducing flame it gives a fine green glass, whose colour increases in beauty by cooling. On charcoal it is difficult to obtain any other than a green colour, although its intensity diminishes in the oxidating flame.

Soda does not dissolve oxide of uranium. With an exceedingly small quantity of this flux, some signs of fusion are perceptible; a larger portion gives the assay a yellowish brown colour, arising from the formation of an oxide, which saturates the alcali like an acid; with a still larger quantity of soda, it is absorbed by the charcoal, but not reduced. Traces of tin are commonly found in the assay if that metal have not been previously separated from the oxide by sulphuretted water,[1] or hydrosulphuret of ammonia.

19. *Oxides of Cerium.*

Alone, the protoxide passes to the state of peroxide. The latter suffers no change even in the reducing flame.

Borax dissolves the oxide in the exterior flame, and forms a beautiful red, or deep orange yellow glass, whose colour fades on cooling, and is ultimately reduced to a yellowish tint; by *flaming*, the glass becomes enamel white. In the reducing flame it loses its colour. With a larger proportion of oxide, in the reducing flame the glass turns spontaneously enamel white, and becomes crystalline on cooling.

With salt of phosphorus the oxide fuses into a fine red glass, which loses its colour on cooling, and becomes as limpid as water. In the reducing flame the glass becomes colourless, but never dissolves enough of the oxide to be opaque on cooling.

With soda, it does not fuse; the soda is absorbed by the charcoal, and white, or greyish white protoxide of cerium remains on the surface.

The effects of oxide of cerium very much resemble those of oxide of iron, especially if the ce-

---
[1] Solution of sulphuretted hydrogen. C.

rium be combined with silica, for that substance prevents the glass which it forms with borax from losing its transparency; the protoxides of iron and cerium, however, do not behave alike with respect to the fluxes; nevertheless, when in simultaneous combination with silica, as they usually are, the presence of oxide of cerium cannot be ascertained by the blowpipe.

20. *Oxide of Manganese.*

Alone, does not fuse, but turns brown in a strong heat.

With borax, fuses into a transparent amethyst coloured glass, which loses its colour in the reducing flame. If the proportion of oxide be large, the instant we discontinue the blast the reduced glass must be thrown on a cold body; it would recover its colour if cooled slowly. With a very large proportion of oxide, in the exterior flame, the glass ultimately assumes so deep a colour, that it appears black; but when drawn out to a thread its transparency is evident.

With salt of phosphorus it fuses easily into a transparent glass, colourless in the reducing flame, and amethyst red in the oxidating, but never so deep as to render the glass opaque. Whilst in fusion in the exterior flame, whether on the platina wire or on charcoal, the glass boils up and gas is disengaged. This effervescence ceases in the reducing flame, but immediately begins again by exposing the glass to the oxidating flame. We may explain this phenomenon, by saying, that the glass globule is peroxidated at the surface; that its rotatory motion carries the peroxide into the interior of the mass, when the oxygen is expelled by the phosphoric acid, at the same time that the salt of peroxide is converted into salt of protoxide. Hence also, the salt of phosphorus never acquires the amethyst colour beyond a certain extent, for the glass can only contain a certain quantity of salt of peroxide. In general, the metallic glass formed with borax is more easily retained in the oxidated than in the reduced state; the glass made with salt of phosphorus, on the contrary, rather preserves a state of complete reduction, and cannot be perfectly oxidated.

If the glass of salt of phosphorus and oxide of manganese, contain so small a portion of the latter that it appears colourless, the colour may be developed by applying a crystal of saltpetre to the fused globule, in the manner described under the article *saltpetre* (page 58). The saltpetre causes the assay to foam up, and the foam on cooling assumes an amethyst or pale rose colour, according to the proportion of the oxide of manganese.

With soda on the platina wire, or foil, oxide of manganese fuses in very small quantity into a green transparent mass, which fixes on cooling, and becomes bluish green. The experiment succeeds best on the foil; for as the oxide of manganese dissolves in the soda, the liquid mass flows over the surface of the platina, whence its colour is easily seen after it is cold. Oxide of manganese, not exceeding 1-1000th part of the weight of the assay, gives a sensible green colour to the soda; by means of this colour the slightest traces of manganese may be detected. Manganese is not reducible by soda on charcoal; but if it contain a very little iron, that metal may be reduced and separated in the usual manner.

21. *Oxide of Zinc.*

Alone, it assumes by heat a yellow colour, distinctly visible by day, but not by candle light. Its white colour returns on cooling. It does not fuse, but gives out a brilliant light whilst in the state of incondescence, and is gradually dissipated before the reducing flame, a white vapour condensing at the same time on the surface of the charcoal.

With borax it fuses readily into a transparent glass, which becomes milky by *flaming*, and with a larger portion of oxide assumes the whiteness of enamel on cooling. In the reducing flame the metal sublimes, and the charcoal is covered with white fumes at a small distance from the glass.

With salt of phosphorus it behaves as with borax, except that it is not so easily reduced and sublimed with the first as with the second. With regard to its re-actions, it resembles most of the earths, properly so called.

Soda does not dissolve it; but when treated with

that reagent on charcoal, it is reduced, and covers the support with vapours of zinc; with a good heat the zinc may even be inflamed. This is the principal character of the oxide of zinc; and in those minerals which contain it, as for instance, gahnite, it is from the white fumes which cover the charcoal, when they are acted on by soda, that we ascertain the presence of this oxide.

With solution of cobalt it gives a green colour.

22. *Oxide of Cadmium.*

Alone, on platina foil, in the exterior flame it undergoes no change. On charcoal it is dissipated in a few seconds, and the charcoal is covered with a red or orange yellow powder. This phenomenon is so marked with oxide of cadmium, that minerals, which, like carbonate of zinc, contain one or two per cent. of carbonate of cadmium, when exposed for a single instant to the reducing flame, deposit, at a little distance from the assay, a yellow or orange coloured ring of the oxide, most distinct when the charcoal is cold. This ring forms long before the oxide of zinc begins to be reduced, and if flocculi of that metal appear at the same time, it is a proof that the blast has been pushed too far; but if we can discover no yellow trace before the fumes of zinc begin to condense on the charcoal, we may conclude that the assay contains no cadmium.[1]

[1] I am indebted for the following to the kindness of my friend, Dr. E. D. Clarke:—If platina foil be used for the

Borax dissolves it in large quantity on the platina wire, and gives a transparent glass of a yellowish colour, which in great measure disappears on cooling; if the glass be *nearly* saturated it becomes milky by flaming, and if *completely* saturated it spontaneously assumes the whiteness of enamel as it congeals. On charcoal it keeps up a continual bubbling, the cadmium is reduced and sublimes, and the charcoal is covered with the yellow pulverulent oxide of cadmium.

Salt of phosphorus fuses it in large quantity into a transparent glass, which, if saturated, becomes milk white on cooling.

With soda, on the platina wire, it does not fuse. On charcoal it is reduced, sublimes and leaves a circular trace of a yellowish colour.

23. *Oxide of Iron.*

Alone, in the exterior flame, no change; but blackens and becomes magnetic in the interior flame.

With borax, in the oxidating flame, it fuses into a dull red glass, which becomes clear on cooling, and finally assumes a yellow tint, or entirely loses

support, instead of charcoal, the yellow or orange coloured ring is not only conspicuous upon the platina, but by exposing it to the point of the blue flame, the metal is revived, and afterwards deposited upon the platina, during its combustion, in the form of a protoxide, exactly like polished bronze or copper. C.

its colour. If the proportion of oxide be large, the glass is opaque whilst fluid, and assumes an impure, dull yellow colour on cooling. In the reducing flame it becomes bottle green, and if the reduction be pushed as far as possible, it acquires a lively bluish green colour, exactly like that of the vitriol obtained by dissolving iron in diluted sulphuric acid. Tin accelerates the total reduction of the peroxide to the state of protoxide. The bottle green colour is derived from a mixture of the peroxide and protoxide,[1] and is sometimes so deep that it appears black. As long as the glass contains only the peroxide of iron, it continues transparent whilst in fusion; but as soon as it is exposed to the reducing flame and the mixed oxide begins to be formed, it becomes opaque, and remains so till the whole is converted into protoxide, when it resumes its transparency. The green colour of the protoxide is very beautiful whilst the glass is hot, but it fades on cooling, and is imperceptible if the quantity of iron be small.

With salt of phosphorus, it fuses and presents the same phenomena as with borax, but the colour fades more on cooling: tin makes it disappear

---

[1] " *Oxide ferroso-ferricum*,"—a compound, according to Berzelius, of one atom of protoxide and two atoms of peroxide. I have given my reasons in the preface for not adopting my author's nomenclature in this and many subsequent instances. I do not expect that he will acquiesce in the change I have adopted, but, as no *disrespect* is intended, I trust he will pardon me. The English reader, I hope, will pardon and *approve*. C.

almost wholly. A glass, containing a large quantity of oxide of iron, gives with tin a pale bluish green colour; sometimes, at the first moment of cooling, the vitrious globule has a superficial pearl grey coating, but this appearance vanishes if we renew the blast.

Soda does not dissolve oxide of iron, but is absorbed with it by the charcoal, where the oxide is easily reduced, and, after the assay has ceased to foam, affords a grey powder which is metallic and magnetic.

24. *Oxide of Cobalt.*

Alone, no change.

With borax, fuses readily into a transparent blue glass, which does not become opaque by flaming. A small quantity of oxide gives a deep coloured glass; with a larger quantity it appears black.

With salt of phosphorus it also fuses readily into a blue glass; by candle light the colour appears violet—by day, pure blue. If the salt of phosphorus have only a slight blue tint by day, it appears rose colour by candle light.

Soda, on the platina wire, dissolves but very little oxide of cobalt; when viewed by transmitted light, the fused mass has a pale red colour, and becomes grey on cooling. On platina foil, the fused portion of the oxide flows over the surface, and envelopes the unfused portion with a thin dull red coating.

With subcarbonate of potassa, oxide of cobalt fuses in much larger quantity, the salt does not flow so much abroad, and the congealed mass is black, without any mixture of red. This might be used as a characteristic to distinguish the two alkalies, (soda and potassa) if it were not quite as easy, and much more certain, to expose the alkali we are in doubt about to the air, and observe if it become moist or not.

Oxide of cobalt is very easily reduced in the interior flame with an alcali, or an alcaline salt on charcoal, though we take so small a portion of the re-agent that the mass is not absorbed; but it does not fuse. After washing away the soda and the charcoal, the residuum is a grey metallic powder, which is magnetic, and assumes under the polisher the characteristic brilliancy of metals.

25. *Oxide of Nickel.*

Alone, no change.

With borax it fuses readily into an orange yellow, or reddish glass, which becomes yellow, or almost colourless, on cooling. With a larger quantity of the oxide the glass is opaque and dull brown whilst in fusion; but, on cooling becomes dull red and transparent, like ferruginous tungstic acid with salt of phosphorus. The reducing flame destroys the colour and the glass becomes grey, from the mixture of finely divided particles of metallic nickel dispersed through the mass. By continuing the blast, these particles may be united, but not fused. If the oxide of nickel contain cobalt, as often happens, the colour of that metal prevails; if arsenic also be present it fuses into a bead.

With salt of phosphorus it fuses with the same phenomena of colour, as with borax; but the colour almost wholly disappears on cooling. It behaves in the same way in the oxidating and reducing flame, whence it is distinguished from oxide of iron, which it otherwise much resembles in its characters before the blowpipe. Tin at first produces no change, but afterwards the nickel is precipitated, and the colour disappears. Cobalt, if present, may then be perceived, but the blue glass it affords is opaque; and, in general, cobalt is not so well detected in this manner as with the glass of borax.

Soda does not dissolve oxide of nickel: a large quantity of this flux causes it to be absorbed by the charcoal, where it is readily reduced; and, by washing, we obtain small, white, brilliant, metallic particles, more strongly attractable by the magnet than, perhaps, even soft iron. If the proportion of soda be small, the mass remains on the surface of the charcoal, but the nickel is nevertheless reduced, and may be collected by washing. Pure nickel is not fusible by the blowpipe. Nickel containing traces of arsenic does not fuse with soda; but if we add borax it fuses into a globule, capable of being flattened under the hammer, but whose edges commonly split, and are magnetic in a high degree.[1]

26. *Bismuth, Oxide of Bismuth.*

Alone, on platina foil, oxide of bismuth fuses readily into a dull brown mass, which becomes yellowish on cooling. With an intense heat, it fuses and perforates the platina. On charcoal, it is instantly reduced into one or several metallic globules.

With borax, in the exterior flame, it fuses into a colourless glass. In the interior flame it is reduced, and gives a greyish glass, which is obscure from the metallic particles dispersed through it.

Salt of phosphorus dissolves it into a yellowish brown glass, whilst hot; and, when cold, colourless, but not perfectly clear. In the reducing flame, particularly with tin, we obtain a glass, clear and colourless whilst hot, but opaque and greyish black when it congeals. Protoxide of copper presents nearly the same phenomena under the same circumstances, except that it gives a red colour. This fact seems to indicate, that there is a lower state of oxidation at which bismuth is capable of acting as a salifiable base than that determined *viâ humidâ.*

[1] " Mais dont les *bords* se déchirent ordinairement sous le marteau, et sont *magnetiques* à un haut degré." Does this mean that the *edges* only are magnetic? I give it as I find it. C.

From the facility with which bismuth is reduced, it happens that in operating on it by the blowpipe, the experiment almost always bears upon the *metal* itself. Hence it becomes very important to know how to distinguish it from antimony and tellurium, with which it may easily be confounded.

(*a.*) *In the matrass* neither antimony nor bismuth sublime at any temperature that the glass can support without fusing. Tellurium, on the contrary, first gives off a few fumes (from the oxygen of the atmospheric air), and afterwards we obtain a grey sublimate of metallic tellurium.

(*b.*) *In the open tube* antimony gives off white fumes, which cover the interior of the tube, and may be driven from one part of it to another without leaving any mark. The metallic globule becomes surrounded by a considerable quantity of fused oxide.

Tellurium gives off a large quantity of fumes, which adhere to the sides of the tube as a white powder, capable of fusion into clear colourless drops. A small portion sublimes, but the whole of the remainder is converted into these little drops, which, however small, roll about the surface of the glass when acted on by the heat. If the layer of white powder be thin, it insensibly disappears during the operation, as if it were sublimed; but, by the microscope, we may perceive that it is converted into exceeding small drops. The metallic globule is surrounded by a clear, almost colourless, fused oxide, which, on cooling, becomes white, opaque, and foliated, in the thicker parts. With a brisk heat, and a gentle current of air, a part of the metallic tellurium sublimes and condenses in the form of a grey powder.

Bismuth (provided it be free from sulphur) gives scarcely any fumes,[1] and the metal is surrounded by a dull brown fused oxide, which retains, when cold, only a yellowish tint. It acts strongly on the glass.

(*c.*) *On charcoal* all the three metals fly off in fumes by a gentle heat, leaving an areola round the spot where they lay. That of antimony is quite white; those of bismuth and tellurium are red or orange at the edges. If the reducing flame be directed on this mark, it disappears, and the flame assumes a fine deep green colour if the areola be owing to tellurium, and a pale greenish blue if to antimony. If the areola be occasioned by bismuth, the flame is not coloured. I should add here, that the odour of putrid horse-radish, attributed to tellurium, is not at all perceptible if the metal be pure; it arises entirely from the selenium, which, in some minerals, accompanies tellurium.

27. *Oxides of Tin.*

Alone, the protoxide, whether pure, or in the state of hydrate, takes fire and burns like tinder, and is converted into peroxide. The peroxide

---

[1] See *Sulphuret of Bismuth*, amongst the minerals. B.

does not fuse nor suffer any change;[1] but is perfectly reducible to the metallic state by a strong reducing flame well kept up without the help of any re-agent: the operation, however, requires skill.

With borax fuses with difficulty and in small quantity into a transparent glass, which remains so on cooling, and cannot be converted into white enamel by flaming; but, if the glass be saturated with oxide, and, when completely cold, it be heated afresh in the exterior flame to incipient redness, it then becomes opaque, loses its round form, and suffers a sort of confused crystallization. The colour of the glass does not change in the reducing flame.

With salt of phosphorus oxide of tin fuses with difficulty and in small quantity into a transparent colourless glass. If we add oxide of iron, it (the oxide of iron) loses the property of colouring the glass, but with certain limitations, for a given quantity of oxide of tin, can deprive only a proportionate quantity of oxide of iron of its colouring power,—the *excess* would impart colour to the glass as if no oxide of tin were present. Arsenic makes the glass opaque.

With soda, on the platina wire, oxide of tin combines with effervescence, forming an infusible turgid mass, insoluble in an additional quantity of soda. On charcoal it is readily reduced into a globule of metallic tin. Some native oxides of tin,

---

[1] It seems as if something were omitted here, probably the words, " in the oxidating flame." C.

especially such as contain columbium, are not easily reduced with soda, so that, on a first experiment, one might doubt its presence; but if we add a small quantity of borax, the reduction is immediately accomplished.

Tin often occurs in nature, as an accidental and, relatively, very small constituent part of the ores of columbium, titanium and uranium, and perhaps of some others, where its presence would be little suspected in experiments made in the moist way; but when treated with soda in the reducing flame, particularly after separating the iron, metallic tin is always detected even though it enter in no larger proportion than 1-200th part of the weight of the ore. If the proportion of iron be inconsiderable, its reduction may be prevented to a certain point, by adding borax to the soda.

28. *Oxide of Lead.*

Alone, minium appears black whilst hot, and at an incipient red heat changes to the yellow oxide. The latter fuses into a fine orange-coloured glass, which on charcoal is reduced with effervescence to a globule of lead.

With borax on the platina wire, it fuses readily into a transparent glass, which, when saturated, is yellow whilst hot, but becomes colourless on cooling. It does not retain its globular form on charcoal, but spreads over the surface, whilst the lead is reduced with bubbling and flows towards the edges.

With salt of phosphorus it fuses easily into a transparent colourless glass. When saturated it appears yellowish whilst liquid, and becomes enamel white on cooling. It is not reduced in the interior flame unless there be an excess of oxide of lead.

With soda on the platina wire oxide of lead fuses readily into a transparent glass, which becomes yellowish and opaque on cooling. On charcoal it is instantly reduced.

29. *Oxide of Copper.*

Alone, in the oxidating flame, oxide of copper fuses into a black globule, which soon spreads over the charcoal, and is reduced at the lower surface. In the reducing flame, and at a temperature not sufficient to fuse copper, the oxide is reduced, and displays the metallic brilliancy of that metal; but, as soon as the blast is suspended, the metallic surface re-oxidates, and becomes black or brown.[1] In

---

[1] Gahn, who had considerable copper works at Fahlun, which he superintended with great care, observed that ores from different parts of the mine required different modes of treatment, that the loss of metal in roasting should not be excessive. To ascertain at once if a scoria contained more copper than it ought, he assayed it by the blowpipe, by exposing large thin scales, first to the oxidating flame, to burn off the sulphur, and then to the reducing flame, in such a manner as to cause it to cover the whole roasted surface. If a scoria contain copper, points, striæ and marks, having the colour and brilliancy of that metal, are perceptible on the surface; and their quantity indicates that of the copper in the slag.

a stronger heat it fuses into a globule of metallic copper.

With borax in the oxidating flame, oxide of copper fuses readily into a fine green glass, which becomes colourless in the reducing flame, but on congealing assumes a colour bordering on cinnabar red, and becomes opaque. If the oxide of copper be impure, the glass usually becomes deep brown, and does not assume the appearance of enamel unless exposed to an intermitting flame. If the proportion of oxide be considerable a part of it is reduced in fused globules, which are obtained by pounding the glass.

With salt of phosphorus it fuses and exhibits the same shades as with borax. If the proportion of copper be inconsiderable, the glass sometimes becomes transparent and ruby red, after exposure to the reducing flame, nearly at the moment of congelation. Commonly the glass becomes red and opaque like enamel.

If the quantity of copper be so small that the character of the protoxide cannot be developed by the reducing flame, we add a little tin to the assay (whether the flux be salt of phosphorus or borax) and immediately continue the blast. The before colourless glass now becomes red and opaque on cooling. If the blast be kept up too long, the copper precipitates in the metallic state, particu-

---

It is seldom that scoriæ are entirely free from copper, but the eye can soon distinguish an excessive, from the ordinary, quantity. B.

larly with salt of phosphorus, and the colour is destroyed.

With soda on the platina wire, oxide of copper fuses into a fine green glass, which loses its colour and transparency on cooling. On charcoal the mass is absorbed and the oxide reduced. There is probably no other possible method of discovering such minute proportions of copper as may be detected by the blowpipe, in all cases where it is not combined with other reducible metals, liable to disguise its properties. In the latter case we must employ borax and tin. If copper and iron be found together, the same operation reduces each separately into distinct particles, which may be known by their respective colours, and separated by the magnet.

30. *Mercury.*

All the combinations of mercury are volatile, and consequently cannot be acted on by fluxes. Substances containing mercury are assayed by mixing them with a little metallic tin, or iron filings, or oxide of lead and heating the mixture to redness in a glass tube closed at one end. The mercury is reduced, and sublimes into the cold part of the tube, in the form of a grey powder, which, by agitation, collects into little metallic globules.

31. *Oxide of Silver.*

Alone it is reduced in an instant.

With borax it is partly dissolved and partly reduced. In the oxidating flame the glass turns, on cooling, milk white or becomes opaline, according to the quantity of silver dissolved, even if the silver have been added in the metallic state. In the reducing flame it assumes a greyish appearance from the particles of reduced silver disseminated through the mass.

With salt of phosphorus both the oxide and the metal, in the oxidating flame, give a yellowish glass, which assumes the colours of opal if the proportion of silver be augmented; seen in daylight, by refraction, it appears yellow; viewed in the same manner by candle-light, it has a reddish colour. It becomes greyish in the reducing flame, like the glass of borax.

The other noble metals, gold, platina, iridium, rhodium and palladium, have no action on the fluxes and are incapable of oxidation. All that can be done with them by those reagents, is to ascertain whether they contain any of the other more oxidable metals, which, combining with the fluxes, colour them. We may also fuse them with perfectly pure lead and cupel them on bone ashes, in order to judge by the colour of the cupel loaded with the oxide of lead, if any foreign metals be present. Of the noble metals above named, gold is the only one that can be obtained in a button; each of the others, after the lead is separated, forms a grey, slightly porous, infusible mass, which, under the steel polisher, assumes metallic brilliancy. Platina and palladium are malleable,

old, 20. O. F. c
colour flies in ery minute portion of manganese
een glass with soda.

with 21. O. F. f
har- comes milky by

cir- 22. P. W.
on cooling; on
reduced, sublim

sed. 23. O. F. du
and yellowish, e reduction of iron from the per-
C. and R rotoxide is facilitated by tin.
bluish-green.

tted 24. Fuses rea

ised. 25. O. F. o
glass; becomes
less, on cooling.

 glass matrass does not sublime
ing point of glass. In an open
cely gives off any fumes; the
mes covered with a dull-brown
a. O. F. colou de, of a slight yellowish tint,
R. F. partly
glass.

nfu- 27. Fuses with
nently clear glass.

28. P. W. c
on cooling, colour
C. flows over th

29. O. F. fin
R. F. becomes co
nabar-red and opa

 the compounds of mercury are
ixed with tin or iron filings, and
a glass tube, metallic mercury
.

31. O. F. glas
opaline, on cooling
R. F. greyish.

hese metals have no action on
fluxes, which can only serve to
ct the foreign metals they may
combined with. They are best
on with lead.

B. SUBSTANCES RESULTING FROM THE COMBINATION OF COMBUSTIBLE BODIES.

1. *Metallic Sulphurets.*

They are known by the odour of sulphurous acid which they exhale when roasted on charcoal, or in a glass tube. If the quantity of sulphur in a metallic compound be too small to render the odour perceptible, we form a bead of glass by fusing silica with soda, on which we place a small particle of the assay; if it contain sulphur, the glass, either immediately or on cooling, assumes a red or yellow colour, according to the proportion of sulphur. But if the colour from the sulphur be concealed by colour derived from the metal, the assay must be roasted in an open tube, containing in its upper part a piece of paper tinged with infusion of brazil wood, which will be bleached by a portion of sulphur quite insensible to the smell. This process should be particularly adopted in roasting the ores of antimony, in which it is difficult to distinguish the odour of sulphur, on account of the equally pungent odour developed by the antimony.

The principal object in assaying metallic sulphurets, is to ascertain the metal combined with the sulphur; wherefore the latter substance must be driven off as completely as possible by roasting. We therefore select thin laminæ of the assay, on which the air has more effect than on equal masses of a more solid figure. That they may retain their lamellar form, we first heat them gently, so as not to fuse them. If fusion inadvertently ensue, it is best to take a fresh assay. At a certain point of the operation, some sulphurets lose the property of fusing; the heat may then be increased to shorten the process and decompose the sulphate, usually formed at its commencement. The roasting succeeds well on charcoal; it must not be attempted on platina foil, because the foil is frequently acted on by the metal. If charcoal be objectionable, the operation may easily be performed on a plate of mica, taking care to select one that is not too fusible.

We cannot, till the roasting is completed, avail ourselves of the effects produced by the fluxes. The reduction by soda requires particularly the entire expulsion of the sulphur; if ever so small a portion be left, either metallic sulphurets, in which the metals cannot be ascertained, are formed, or the metals are dissolved and carried off by the sulphuret of soda, so that nothing remains in the mortar after washing the mass.

2. *Metallic Seleniurets.*

These are more readily known than any of the other metallic compounds of the same order, by the odour they emit when heated in the exterior flame; the better to distinguish which, the assay must be applied to the nostrils whilst hot. This odour is very strong and very disagreeable, and resembles that of decayed horse radish. The smallest portion of selenium may be detected by it.

With glass of silica and soda, the seleniurets afford the same phenomena as the metallic sulphurets, but their colour disappears more readily by a prolonged blast, than that of the latter.

Selenium may often be easily obtained in the metallic state,[1] by roasting in an open tube. By inclining the tube as occasion requires, we may so regulate the current of air as to oxidate the metals combined with it, whilst the selenium sublimes and displays a red colour. If a seleniuret and a sulphuret be found together, the selenium sublimes in its elementary state, whilst the sulphur is disengaged in that of sulphurous acid. Some Swedish galenas contain a small portion of selenium, which may be detected in this way. If selenium be mixed with tellurium, oxide of tellurium first sublimes, and afterwards, nearer to the heated spot, selenium is deposited in the form of a red powder. Sulphuret of arsenic sometimes sublimes with all the appearances of selenium, but its odour is different.

3. *Alloys of Arsenic.*

Arsenic is detected by its odour, when heated. We must here remember, that it is not *aresenious acid*, but *metallic arsenic* in its volatile state, that emits the odour of garlick. If the arsenic be in

---

[1] If selenium must be a metal, it must; but surely its characters do not warrant our considering it as one. It's not being a conductor of heat or electricity, seem to me decisive. C.

large proportion, the assay fumes abundantly, and the odour is perceptible at a considerable distance; when its quantity is less, after the assay has been exposed to a good reducing flame, it must be brought near the nose whilst red hot; if the proportion be very small, its odour cannot be discovered till the assay has been treated with soda in the reducing flame. The odour of arsenic is so good a character, that it may even be detected by it in the small portion of smalt, commonly used to give a blue tinge to paper, by exposing the ashes of the paper to the reducing flame. In roasting arsenical alloys, the operation is best begun in a glass tube, to the sides of which the greater part of the arsenic will attach itself, in the form of a white crystalline sublimate of arsenious acid, instead of diffusing itself through the air. There is also another advantage in this method—namely, that the odour of the sulphurous acid, if any be present, is more perceptible after the gas has deposited its arsenic on the glass. When the greater part of the arsenic is thus separated, the roasting is completed on charcoal, using alternately the oxidating and the reducing flame; because one part of the arsenic combines, as an acid, with the metallic oxides, and must be brought back to the state of metal by the reducing flame, in order to roast it afresh by the oxidating flame. It is, perhaps, even more necessary to expel all the arsenic from an alloy, than the sulphur from a sulphuret, especially in experiments of reduction; for metals containing arsenic are ascertained with greater difficulty than those combined with sulphur.

In roasting arsenical minerals, care must be taken not to expose oneself unnecessarily to the vapours, which are always dangerous. I confess, however, that I have often been in chambers, the air of which was loaded with the smell of arsenic, without having ever experienced any ill consequences, and I have been astonished at seeing the workmen at the silver foundries near Freyberg, daily immersed in an arsenical atmosphere, without their health appearing to be injured by it.

4. *Alloys of Antimony.*

When roasted in an open tube they give off fumes of antimony, the nature of which varies according to the metals with which the antimony is combined. If they be very oxidable, a large portion of the antimony passes off in the state of antimonious acid, and its vapour is infusible and fixed; but if the antimony be combined with copper or silver, a volatile sublimate of oxide of antimony condenses on the glass; the vapour which escapes from the tube has a pungent, but not fœtid odour, arising either from the oxide of antimony, as it seems, or from the antimonious acid.

5. *Alloys of Tellurium.*

Roasted in the glass tube, they deposit on its upper part the same pulverulent coating as pure tellurium. (See Bismuth, p. 111.) The vapour that escapes from the tube has a pungent odour, analagous to that of the vapour of antimony; if it have the smell of horse-radish, it is a proof that the assay also contains selenium. The pulverulent coating, produced by the oxide of tellurium, is distinguished from that of arsenious acid, in being fusible, and *not* crystalline, whereas the latter *is* crystalline, and volatilises without fusing.

6. *Metallic Carburets.*

Metallic carburets, analogous to the metallic sulphurets and alloys of arsenic, those, namely, that are convertible into carbonates, are not found in the mineral kingdom. They may be formed artificially by the dry distillation of certain salts composed of metallic oxides and vegetable acids, or by heating different metallic cyanurets to redness in close vessels; they have the combustibility of charcoal, take fire like tinder, and, in burning, liberate their oxides. But these properties are chiefly owing to their loose texture. The metallic carburets most frequently found in nature, are supersaturated with carbon; graphite is a familiar instance. The only characters by which carbon can be distinguished in the dry way, are the following:—It burns away gradually without either smell or smoke; pulverised and mixed with saltpetre, it detonates on platina foil and leaves a residuum of carbonate of potassa.

C. CHARACTERISTIC EFFECTS OF ACIDS, AS COMPONENT PARTS OF SALTS.

From what has been said on the phenomena presented by the metallic oxides, separately considered, we are now able to ascertain the nature of the base of any saline compound, particularly if it be metallic. It remains to show how the acid may be known.

1. *Sulphuric acid* is detected by placing an extremely small quantity of the salt to be assayed on a globule of silica and soda fused together, or by mixing the salt with the soda previous to its fusion with the silica. The second process is the easiest, but the first is most certain. The sulphuric acid is reduced in the operation, and sulphuret of soda is formed; the glass immediately assumes a dark brown tint, or becomes colourless whilst in fusion, and red or orange on cooling, according to the quantity of salt operated on.

2. *The Nitrates*, if fusible, detonate with charcoal; those which do not fuse are to be heated, at first to perfect dryness, and then red hot, in a glass tube closed at one end: the tube is soon filled with the orange yellow vapour of nitrous acid.

3. *Muriatic acid.*—I had in vain attempted to find a proper test for this acid in the dry way, when I was led to the following by an observation

of Bergman's; it succeeded beyond my hopes.[1] We fuse oxide of copper with salt of phosphorus into a dark green globule; we then add the assay, and heat the whole before the blowpipe. If it contain muriatic acid, the globule is surrounded by a fine blue flame, inclining to purple, which continues as long as any muriatic acid remains in the assay. Not one of the other mineral acids produces a similar phenomenon, and such of them as form cupreous salts which do of themselves colour the blowpipe flame, lose that property when combined with salt of phosphorus. For instance, the earthy mineral, in which the blue carbonate of copper (from Chessy, in France,) occurs, communicates an intense green colour to the flame when heated before the blowpipe; but when treated with salt of phosphorus, previously saturated with oxide of copper, not the slightest colour is any longer perceptible in the flame.

4. *Iodates.*—These, treated in the same manner as the muriates, colour the flame with a superb deep green. It is well to observe in this place, that the salt of phosphorus, as it begins to fuse, sometimes throws out little pale green flames, arising from the combustion of the ammonia; and, when heated pretty strongly in the reducing

---

[1] Bergman observed, that muriate of copper colours flame green—an effect which no salt of copper, formed with the other mineral acids, produces. B.

flame, it again produces a greenish hue; but these appearances cannot be confounded with the brilliant green flame developed by an iodate.

5. *Fluates.*—Since the discovery of fluoric acid in many minerals where its presence was not heretofore suspected, as in wavellite, hornblende and mica, it is become important to have the means of detecting it by the blowpipe. This is more difficult with those compounds of which fluoric acid is an essential part, as fluor spar, topaz, cryolite, &c., than where it seems only to be an accidental ingredient, as in mica and hornblende. From the first, fluoric acid cannot be expelled by heat; from the second, the relative position of their constituent parts changing at the moment of ignition, the fluoric acid is disengaged, and usually carries off with it a certain quantity of silica.

If a mineral be very rich in fluoric acid, we mix it with salt of phosphorus, previously fused, and heat the mixture in an open tube, placing it near its lower end, so that a part of the current of air which feeds the flame may enter the tube: fluoric acid vapour is thus formed, which fills the tube, and may be known both by its peculiar odour and by its corroding the glass, which becomes dull through its whole length, and particularly at those places where the vapour condenses. If paper stained red with infusion of brazil wood, be held to the mouth of the tube, it immediately turns yellow.[1]

---

[1] According to Bonsdorff's experiments, the fluoric, phos-

When, on the contrary, fluoric acid exists in a mineral in small quantity, in combination with a weak base and a minute proportion of water, we need only heat the assay in a tube closed at one end, and containing a small slip of moistened brazil wood paper. The heat disengages the silicated fluoric acid, an unpolished ring of silica forms on the glass at a little distance from the assay, and the end of the slip of paper becomes yellow; sufficient indications of disengaged fluoric acid. In this manner its presence may be detected in that species of mica which contains only about 1-130th of its weight of fluoric acid.

6. *Phosphates.* The unexpected discovery of phosphoric acid in wavellite and lazulite, has shown the necessity of a test for this acid, the rather, since the property of being precipitated with the earthy bases, often occasions it to elude the investigation of the chemist in experiments performed in the moist way. Reflecting on the known phenomena presented by phosphate of lead,

---

phoric and oxalic acids turn brazil wood paper straw-yellow, a property not possessed by the sulphuric, nitric, muriatic, boracic, &c. acids. But he observed that some of the latter, when diluted with a certain quantity of water, also give it a yellowish tinge, though less lively, and slowly developed; whereas the effect of fluoric acid is immediate, and the colour a fine yellow. A fluate may often be ascertained, by merely moistening it in a watch-glass with muriatic acid, and, after a few seconds, rubbing a piece of brazil wood paper with the mixture, when the peculiar effect of fluoric acid will be evident. B

I endeavoured to find a method of detecting it by means of lead or its oxide; but my researches were in vain as to all the phosphates, except phosphate of copper; I could not accomplish my object without previously combining the phosphoric acid with the oxide of lead in the moist way,—a process inadmissible in the series of experiments I had undertaken. After many other fruitless attempts, I at length discovered the following effectual method.—We fuse the assay with boracic acid, and, when the fusion is complete, we plunge the end of a little steel wire, rather longer than the diameter of the globule, into it, and heat it in a good reducing flame. The iron becomes oxidated at the expence of the phosphoric acid, whence borate of protoxide of iron and phosphuret of iron result; the latter fuses at a pretty high temperature, and, at the same time, the assay, which had spread itself over the whole length of the wire, resumes the globular form: as the globule cools, an appearance of ignition is generally seen near its base, arising from the crystallization of the phosphuret of iron. We then remove the globule from the charcoal, wrap it in a piece of paper, and strike it gently on the anvil with the hammer to separate the phosphuret of iron, which we find in the form of a brittle metallic globule, attractable by the magnet and having a steel-coloured fracture. Its brittleness depends on the proportion of the iron; it may sometimes be a little flattened under the hammer. If the assay contain no phosphoric acid, the steel wire will burn only at the ends which project beyond the globule, preserving elsewhere its form and brilliancy. Since four or five per cent. of phosphoric acid are insufficient to fuse a mass of iron as large as the experiment requires, to be unequivocal, a proportion not exceeding that quantity cannot be discovered by this process.

Before we proceed to detect the phosphoric acid, we must ascertain whether the assay may not contain some other substance reducible by the iron and capable of being fused with it into a globule, as sulphuric, or arsenic acid, or metallic oxides; since, in that case, we should obtain compounds of their bases and iron.

7. *Carbonates.*—The dry way furnishes no test that can be substituted with advantage for the common one of a drop of muriatic or nitric acid.

8. *Boracic acid.*—I have not hitherto succeeded in my attempts to discover a test for this acid by the blowpipe—a thing much wanted, since, as well as the fluoric, it often occurs in minerals in very small proportion, and frequently escapes detection in analyses made in the moist way.

9. *Hydrates* are easily known by heating the assay in a matrass, when the slightest trace of water will condense in the neck; almost all substances give off a portion.

10. *Silicates* are decomposed by salt of phosphorus, the silica is disengaged, and the base combines with the phosphoric acid. If we employ only a small quantity of the salt of phosphorus, the

silica generally swells up at the moment of decomposition, and absorbs the liquified mass; by a larger portion of the flux, the whole may be changed into a globule, holding in suspension the tumefied semi-transparent silica, which is more distinctly seen whilst the glass is ignited than after it is cool. Most of the silicates give a glass, which is transparent whilst in fusion, and becomes opaline on cooling. If the assay contain but little silica, it generally dissolves entirely in the flux.

Every earthy or stony substance which, with soda, fuses with effervescence into a transparent glass that retains its transparency on cooling, is either silica, or a silicate in which the oxygen of the silica is, in general, to the oxygen of the base, as two (at least) to one. The glass of silica and soda has the property, therefore, of dissolving as much of the base as the soda takes from the silica. But if the assay contain only a small quantity of silica, if, for instance, the quantities of oxygen in the silica and base be equal, the decomposition of the silicate and the formation of the glass take place indeed, but the quantity formed is no longer sufficient to dissolve the base, whose pores absorb the glass. A phenomenon often occurs, in that case, which seems a perfect paradox—namely, that a mineral may form a transparent glass with a very small quantity of soda, which, with a little more, becomes opaque; and, with a still larger quantity, infusible. This phenomenon commonly occurs with the fusible silicates, whose bases, although infusible alone, form, as well as soda, a glass with silica. A small quantity of soda displaces a small quantity of the infusible base, which, however, still remains in solution; every successive addition of soda liberates a fresh portion of the base, and the mass thickens and swells up more and more.[1]

This relation between the phenomena resulting from different proportions of silica has no exception, and always occurs with silicates having the same base; but different bases behave differently in this respect. As the silicates, for the most part, are double salts, whose bases are often combined in

---

[1] In other words, suppose an atom of any base, capable of forming a glass with silica, but infusible by itself, be united in a mineral with an atom of silica; on heating the compound with soda, the result will be a fusible, triple compound, for " the glass of soda has the property of dissolving the quantity of base which the soda takes from the silica," and the globule will be transparent. If two or more atoms of silica be united with one atom of the same base, it will still give a transparent glass; but if there be two atoms of base to one of silica, or the base be in excess in any other proportion, it will afford a result with soda, more or less opaque and infusible, according to the more or less perfect abstraction of the silica from the base by the superior affinity of the soda, and to the quantity of base; provided always, that the base be incapable of forming a fusible transparent compound with the soda. The explanation in the text is founded on views peculiar to the author, by which he has endeavoured to establish as a chemical canon, that, " In combinations of two oxidated bodies, the relation between them is always such, that the oxygen of one of them is a multiple by one, (equal to) two, three, &c.; that is to say, by a whole number of the oxygen of the other."—(*Essai sur la Theorie des Proportions Chimiques.* Paris, 1819, p. 34.) C.

unequal proportions, it happens that two bases, which, in a certain proportion, readily form a glass with soda, in another produce it with great difficulty.[1]

11. The *seleniates, arseniates, molybdates, tungstates* and *chromates,* and those compounds, in which the oxides of titanium and columbium play the part of acids, are known by the characters already described. Thus the seleniates and arseniates are recognised by the odour they emit in the reducing flame, the same as the metallic seleniurets and the alloys of arsenic; the others are sufficiently marked by the effects produced by their acids, which have been individually described.

---

[1] Here follow some illustrations, couched in the theoretical language, and perplexed with the mineralogical signs, peculiar to our author, which I have thought it better to omit, as I shall do all the similar signs that occur in the following description of the action of the blowpipe on minerals; substituting for them, when I may find it necessary, the composition of the mineral in words at full length. (*See preface.*) C.

DESCRIPTION

OF THE

PHENOMENA PRESENTED BY MINERALS

WHEN EXPOSED TO THE

ACTION OF THE BLOWPIPE.

ORDER I.—METALLOIDS.[1]

Of the minerals belonging to this order, there is only one that we have occasion to examine with the blowpipe.

---

[1] Although I have not scrupled to abridge some things pretty freely (especially in the description of apparatus) in the preceding pages, I should not think myself justified in disturbing the author's arrangement in the classification of minerals, in the remaining part of this work. I give it, therefore, exactly as it stands, omitting only the signs, and adding in the notes an explanation of the meaning he attaches to each term, by which his mineralogical orders are distinguished. Thus, *metalloides* are " simple combustible bodies, not possessing the principal characters of metals; as sulphur, boron, and carbon." Nouveau Système, p. 183; to which work I refer the reader for a general developement of our author's views on the subject. C.

Boracic acid from Sasso, in Tuscany. (Placed on moistened brazil wood paper, it bleaches it in the course of half an hour, and turns Turmeric paper, wet with alcohol, brown).

Alone, on charcoal, fuses into a transparent glass. If it contain gypsum, the glass becomes opaque on cooling.

ORDER II.—ELECTRO-NEGATIVE METALS.[1]

1. *Arsenic.*

1. *Native arsenic,* from Saxony.

When heated, gives out the smell of garlick. In the matrass, metallic arsenic sublimes and leaves a small bead of silver.

Observation.—Many species called native arsenic in various collections, are either bi-arseniurets, or mixtures of bi-arseniurets and the metal; for instance, the scapiform arsenic (*stänglicher arsenik*) from Schneeberg, and testaceous cobalt (*Scherben kobolt*), from Saxony, are bi-arseniurets of cobalt.

2. *Sulphuret of arsenic, red* and *yellow.*[2]

---

[1] "Metals whose oxides rather act as acids than bases, in the compounds they form with other oxidated substances." Système, p. 183. C.

[2] *Realgar and orpiment.*—The first composed of 2 atoms of sulphur, 32, and 1 atom of arsenic, 38 = 70; the second of 3 atoms of sulphur, 48, and 1 atom of arsenic, 38 = 86. The

Alone, on charcoal, burns with a pale yellow flame.

In the open tube, burns and deposits white arsenic on the upper part of the tube; it evaporates without leaving any residuum.

In the matrass, fuses, boils up, and gives a dark yellow transparent sublimate; sometimes the sublimate has a fine red colour.

3. *White arsenic.*[1]

Alone, in the reducing flame it gives off the smell of garlick; in the oxidating flame it evaporates without any residuum.

In the matrass it sublimes without previous fusion; the sublimate is crystalline.

2. *Chromium.*

1. *Earthy chrome.* (Chromockra). A mechanical mixture of green oxide of chromium, with quartz and

proportionate weights of the ingredients of those minerals that are unequivocally chemical compounds, as sulphurets, salts, &c. are given from the table at the end of the fourth volume of Thomson's Chemistry, sixth edition, assuming hydrogen as unity, the number of atoms of each substance corresponding to that denoted by the author's signs. The reader will have no difficulty in reducing these weights to parts per cent, by the common rule of proportion. The analyses of the other minerals, whose composition is not so obviously chemical, are quoted chiefly from Phillips's Introduction to Mineralogy, and Jameson's System of Mineralogy. C.

[1] *Arsenious acid,* 1 atom of arsenic 38, and 3 atoms of oxygen, $24 = 62$. C.

transition minerals; from the department of the Saone et Loire.

Alone, it loses its colour and becomes almost white; it does not fuse, but its surface has a scoriaceous appearance, which the microscope shews us is formed of vitrified particles and particles not fused.

Borax gives a fine green glass, with oxide of chrome; the nucleus becomes white and fuses with great difficulty.

Salt of phosphorus, and the oxide, in equal proportions, give a similar glass, but the colour is less intense. The solution is very difficult.

Soda in large quantity ultimately dissolves the earthy chrome. The glass is opaque even whilst liquid, and on cooling resembles a dirty yellowish grey enamel.

Observation.—The earthly chrome of Elfdalen,[1] which appears to lye in spathose albite, behaves in a similar manner, except the difference arising from the nature of its matrix. The same may be said of the chromiferous clay of Mortanberg, only that its whole mass fuses in a strong heat into a black scoria.

3. *Molybdena.*

1. *Sulphuret of Molybdena.*[2]

---

[1] A porphyry quarry in Dalecarlia. B.

[2] Composed of 1 atom of molybdena, 48, and 2 atoms of sulphur, $32 = 80$. C.

Alone, on charcoal, it emits the smell of sulphurous acid, fumes, and leaves a pulverulent deposit on the surface of the support, particularly at the beginning; it burns with great difficulty, its central parts resisting a long continued blast.

With saltpetre, it detonates and fulminates in the spoon, dissolves in the fused salt, and leaves a residuum composed of yellow flakes, which may be obtained separately, by washing away the salt; they behave before the blowpipe like molybdate of iron.

In the open tube gives no sublimate, but the glass close to the assay becomes obscure.

2. *Molybdic acid*, in the form of a very light yellow coating on sulphuret of molybdena. Its characters resemble those of pure molybdic acid; but, treated with soda, it sinks into the charcoal, leaving a residuum of protoxide of iron on the surface.

4. *Antimony.*

1. *Native antimony* from Sala.

Behaves like pure antimony, and is dissipated in fumes without leaving any residuum.

2. *Sulphuret of antimony* black and red.[1]

---

[1] Black sulphuret, composed of 1 atom of antimony, 45, and 3 atoms of sulphur, 48 = 93; the red *calculated* by Berzelius, to consist of 1 atom of oxide of antimony 69, and 2 atoms of sulphuret, 186 = 255. He considers that antimony

Alone, fuses readily on charcoal, which absorbs it and at the same time becomes covered with a black vitreous mass. After blowing a few seconds, metallic globules, apparently a subsulphuret, form in the charcoal, which do not burn like the pure metal, but blacken and tarnish at the surface before they are cold.

In the glass tube, it first gives by roasting a large quantity of antimonious acid; then a mixture of antimonious acid and oxide of antimony (the latter abundant) sublimes. This phenomenon is very remarkable, for the pure metal gives only oxide, and the sublimate is wholly volatile. The air which issues from the tube smells of sulphurous acid.

3. *Alloy of arsenic and antimony*,[1] testaceous antimony from Poullaouen.

Alone, in the matrass, it first gives off much metallic arsenic, then fuses and ceases to sublime. If we remove the globule of metal, and heat it to redness on charcoal, it burns with the same phenomena as antimony, but the fumes have a strong odour of arsenic. Lastly, these fumes crystallize round the metal, but the crystals are whiter and in larger plates, than those of pure antimony; by

---

combines in three proportions with oxygen—the first, oxide of antimony, gives 1 of antimony $+3$ of oxygen,—the second, antimonious acid, $1+4$, the third, antimonic acid, $1+5$. See tables, Nouveau Système, p. 194, or his "*Essai sur la Theorie des Proportions Chimiques.*" C.

[1] One atom of antimony, 45, and 2 atoms of arsenic, $76 = 121$? C.

a prolonged blast the whole of the assay is dissipated in fumes.

4. *Crystallized oxide of antimony* behaves in every respect like pure oxide of antimony.

Antimonious acid derived from the sulphuret of antimony.

Alone, in the matrass, it gives off water; it is, therefore, an aqueous acid. It is not reducible on charcoal, but a slight sublimate of antimony rises.

With soda, it is reduced to metallic antimony. By collecting the globules and subliming them on charcoal we find if the antimonious acid be pure or not.

5. *Titanium.*

1. *Anatase* from Oisans.

It behaves like perfectly pure oxide of titanium. We may remark, that in general the native oxides of titanium dissolve with difficulty in salt of phosphorus, and that the portion not fused becomes white, semi-transparent, and has the appearance of a salt mixed with the mineral.

2. *Rutilite and acicular titanite* behave like oxide of titanium, but the hyacinth colour they give in the oxidating flame, is never so pure as that of anatase. Treated with soda on platina foil, they colour the edges of the flux green,—a proof of the presence of manganese.

Remark.—The rutilite of Käringbricka sometimes give a chrome-green glass with the fluxes, in the oxidating flame; fused with soda on the platina foil, it takes a yellowish colour in the same flame. Sometimes, however, we can obtain no evidence of its containing chrome; it seems, therefore, to be an accidental and variable ingredient.

6. *Silicium.*

Silica in all its forms; as rock crystal, quartz, agate, flint, calcedony, carnelian, &c.

I shall not here describe the effects peculiar to each of these numerous varieties, in which sometimes small qualities of metal produce differences in colour. Their general habits are those of silica, which have been described already. Some varieties, as opal and resinite, also give off water when heated alone in the matrass. But this appears to be merely hygrometric water, similar to that in the dried portions of silica obtained in the analysis of certain minerals, and which varies with the state of the atmosphere.

ORDER III.—ELECTRO-POSITIVE METALS.[1]

DIVISION I.—METALS PROPERLY SO CALLED.

1. *Iridium.*

Alloy of osmium and Iridium. In the form of

---

[1] *Electro-positive metals.* Métals whose oxides act rather as bases then as acids. Système, p. 183. C.

white spangles, collected from the grains of native platina. (Furnished by Dr. Wollaston.) This alloy suffers no change, whether it be treated alone or with the fluxes: having exposed it to a strong heat in the open tube, I thought I perceived the smell of oxide of osmium; but it was too faint to be set down as a characteristic effect.

2. *Platina.*

Grains of platina.—They suffer no change alone, or with fluxes. The phenomena derived from foreign substances mixed with the grains, are not to our present purpose.

3. *Gold.*

1. *Graphic gold*, (Schrift-erz), from Nagyag.[1]

Alone, on charcoal, fuses into a dark grey metallic globule; covers the charcoal with white fumes, that disappear before the reducing flame, giving out a green or bluish light. By a prolonged blast we obtain a clear yellow metallic particle, which at the instant it solidifies becomes for a moment red, or even white hot. When cold it is very brilliant and malleable.

In the open tube it deposits fumes which are white every where except close to the assay, where

---

[1] "It consists of tellurium 60, gold 30, silver 10. *Klaproth.*"—(*Phillips*). C.

they are grey. They consist of sublimed tellurium. These fumes are converted into liquid globules when the flame is directed on them. They have a pungent odour, but not at all similar to that of putrid horse-radish.

2. *Telluriferous and plumbiferous gold.* (Blätter-erz), from the same place.[1]

Alone, on charcoal, it fumes like the preceding, and forms a pulverulent deposit on the support, but the powder is yellow, and is dissipated by the interior flame, giving out a blue colour instead of green. Lastly, in a strong blast, it affords a particle of gold, which ignites at the moment it solidifies, and is malleable.

In the tube it fumes, diffuses a very perceptible odour of sulphurous acid, but not the slightest smell of horse-radish, and gives a sublimate which is grey about the upper part of the assay, and white every where else. The portion of sublimate next the assay does not fuse like oxide of tellurium, but only alters its appearance, and forms a greyish semi-fused covering on the glass, in which we cannot perceive any liquid globules. If the presence of tellurium were unsuspected, we might very easily mistake this portion of the sublimate for antimonious acid; nevertheless, the substance in question is not so white as that acid, which be-

---

[1] *Tellure natif auro-plombifère.* Hauy. *Black tellurium,* Phillips. Tellurium 32·2, lead 54, gold 9, silver 0·5, copper 1·3, sulphur 3. Klaproth. (Phillips). C.

sides has this difference, that it does not become grey and half fused by the effect of heat. The portion of sublimate we are speaking of is tellurate of lead. Farther from the assay, the sublimate has the fusibility and all the characters of oxide of tellurium. The metallic globule attached to the side of the tube, is surrounded by a dark brown oxide, that might be taken for oxide of bismuth, but that its colour scarcely undergoes any change on cooling.

4. *Mercury.*

1. *Cinnabar.*[1]

(*a*). *Crystallized cinnabar*, from Almaden, in Spain.

Alone, on charcoal, it sublimes without leaving any residuum, and exhales the odour of sulphurous acid.

In the matrass it sublimes; the sublimate is blackish, its streak is red.

In the open tube it gives, by roasting, metallic mercury and a sublimate of cinnabar. The mercury condenses farther from the heat than the cinnabar.

In the matrass with soda we obtain globules of mercury.

(*b*). *Mealy cinnabar,*[2] from Zweibrücken.

---

[1] One atom of mercury, 200 + 2 atoms of sulphur, 32 = 232. C.

[2] Native vermilion? C.

Alone, in the matrass, it gives off a little cinnabar, and leaves a considerable residuum, in which the fluxes detect the presence of a large quantity of iron, as well as some lead, and traces of copper.

(*c*). *Hepatic mercury* (Leber-erz).

Alone, in the matrass, it gives cinnabar, and a black residuum. If the latter be burnt in the open tube, it gradually disappears without forming any sublimate and without smell, leaving a little earthy ash. The portion not volatile is therefore analogous to charcoal.

2. *Chloride of mercury.*[1] Horn-erz, from Almaden.

On charcoal, sublimes, leaving only the matrix that may happen to be mixed with it.

In the matrass, gives a white sublimate.

With soda, in the matrass, gives a large quantity of globules of mercury,

With cupreous salt of phosphorus, on charcoal, it tinges the flame a fine azure colour.

5. *Palladium.*

Native Palladium, from Brazil.[2]

---

[1] *Muriate d'oxidule de mercure.* I need make no apology for changing this name. It consists of 1 atom of mercury, 200 + 1 atom of chlorine, 36 = 236. C.

[2] These trials were made on palladium that had not been hammered, and was obtained by reducing the oxide. I know no one but Dr. Wollaston who has had an opportunity of seeing and examining native palladium. B.

Carefully heated to incipient redness by the spirit lamp on platina foil, its surface assumes a blue tint, which disappears on full ignition.

Alone, on charcoal, it is infusible, and unalterable. It fuses with sulphur in the reducing flame. In the oxidating flame the sulphur burns off, and leaves the palladium pure.

6. *Silver.*

Glanzerz. **Sulphuret of silver**, from Schemnitz.[1]

Alone, on charcoal, fuses and swells up considerably, forming empty bubbles, but after the blast has been continued for some time, it collects into a globule. It exhales the odour of sulphurous acid, and at last gives a particle of silver surrounded by scoriæ, which, when fused with borax and salt of phosphorus, give traces of iron and copper.

2. **Red silver**, (Rothgülden) crystallized.[2]

Alone, on charcoal, decrepitates a little, fuses, burns and fumes like antimony, but does not give off any smell of arsenic: the production of the fumes only lasts a few seconds.

In the open tube fumes much and exhales the odour of sulphurous acid, which is particularly sensible at the beginning of the experiment. The

---

[1] Composed of 1 atom of silver, 110+2 atoms of sulphur, 32 = 142. C.

[2] Composed, according to our author's calculation, from Bonsdorff's analysis, of 2 atoms of sulphuret of antimony, 186+3 atoms of sulphuret of silver, 426 = 612. C.

+ refer to XIV

vapour condenses in great measure in the tube, and sometimes forms crystals; it is oxide of antimony, and may be wholly driven off by heat. The globule that remains, after being exposed for some time to the exterior flame, gives a button of pure silver.

3. *Brittle sulphuret of silver* (Spröd Glanzerz) from Saxony.

Alone, in the open tube, fuses, fumes but little, and deposits small white brilliant crystals of arsenious acid on the sides of the glass, without any trace of antimonial vapours.

On charcoal, forms no deposit; takes a long time in roasting; diffuses, with a good heat, a slight odour of arsenic, and parts with its sulphur with much greater difficulty than the glanzerz: it gives a dull grey metallic globule, which is capable of considerable distension under the hammer, but apt to split on the edges. If it be treated in this state with the glass of silica and soda, the glass assumes the colour of sulphuret of soda, and the silver remains pure.

With soda, the roasting and purifying of the silver is accelerated.

The silver may, however, be obtained pure, without the assistance of soda, by a good oxidating flame; whence it is obvious that the foreign substance contained in the assay is volatile. With the fluxes only the characteristic effects of silver are produced.

Remark.—There is so striking a difference in the force with which silver retains the sulphur in the spröd glanzerz, and the glanzerz, that we see, *à priori*, it is owing to the presence of a third substance. According to Klaproth, there are ten per cent. of antimony in the spröd glanzerz from Saxony. I have not found the slightest trace of it; we learn, on the contrary, from experiment, that the presence of antimony facilitates, in the highest degree, the separation of the sulphur, and we obtain, in that case, antimonial silver. But, on fusing together silver and sulphuret of arsenic, I obtained a compound having all the properties of the spröd glanzerz. Hence, I do not hesitate to consider this ore as a compound of sulphuret of silver, with an alloy of silver and arsenic; and, to the latter substance, I ascribe the difficulty of burning off the sulphur.

4. *Antimonial silver* (Spiesglanzsilber) *and argentiferous antimony* (Silber-spiesglanz).[1]

Alone, on charcoal, fuses easily into a grey, brittle metallic globule; gives off fumes like those of pure antimony, but less abundant; after a certain portion of the antimony is driven off, the surface of the globule becomes dull, white, strongly crystalline, and it ignites at the instant it solidifies. After losing a further portion of antimony, its surface becomes smooth, like glass similarly circumstanced, and the heat it then disengages is stronger than at any other moment. Lastly, after a prolonged blast,

---

[1] The first, composed of 2 atoms of silver, 220, + 1 atom of antimony, 45 = 265; the second, of 3 atoms of silver, 330 + 1 of antimony, 45 = 375. C.

nothing remains but silver. During the operation a great quantity of antimonial vapour is deposited on the charcoal, and is sometimes reddish in the direction of the flame, probably from some sulphurous ingredient, which occasions the formation of crocus of antimony.

In the tube much oxide of antimony sublimes, and the residual particle becomes surrounded by a dark yellow glass ring.

5. *Electrum*, composed of gold and silver, probably in variable proportions.

Fuses into a more or less pale yellow globule, which, with borax and salt of phosphorus, presents the same phenomena as pure silver. (See Silver.)

6. *Native amalgam*, from Zweibrücken.[1]

In the matrass, boils up, sputters, and gives off mercury; the residuum consists of a slightly tumified mass, which fuses on charcoal into a globule of silver.

7. *Chloride of silver.*[2]

Alone, on charcoal, fuses into a bead, whose colour, according to the purity of the assay, is pearl-grey, brownish, or black and scoriaceous. In the reducing flame it is gradually converted into a globule of metallic silver.

Chloride of silver is fusible with salt of phos-

---

[1] Composed of 1 atom of silver, 110+2 atoms of Mercury 400 = 510. C.

[2] I have taken the same liberty here as with the author's muriate of mercury. This substance is composed of 1 atom of silver 110+1 atom of chlorine, 36 = 146. C.

phorus, and, if that reagent have been previously mixed with oxide of copper, a sky blue aureola may be perceived playing round the metallic globule: that produced by chloride of mercury, under the same circumstances, has a brighter colour.

7. *Bismuth*.

1. *Native bismuth*, from Schneeberg.

Alone, fuses, giving off a slight odour of arsenic; in other respects it presents the same phenomena as pure bismuth.

In the open tube gives off a little white arsenic. When cupelled, it tinges the bone ashes pure orange yellow.

2. *Sulphuret of bismuth.*

a. *Sulphuret of bismuth, called native bismuth, from Bispberg.*

Alone, in the tube, gives off sulphurous acid and a white sublimate; heated to redness, it boils up, and immediately subsides again; deposits oxide of bismuth on the sides of the tube and round the assay, like pure bismuth.

On charcoal, fuses, boils up, and for a short time projects little red hot drops. After the separation of the bismuth there remains a small quantity of scoriæ, which, fused with salt of phosphorus, give the tinge of iron.

b. *Sulphuret of bismuth*, from Riddarhytta.[1]

[1] Composed of 1 atom of bismuth 71 + 2 of sulphur, 32 = 103. C.

In the tube, at first a little sulphur sublimes, then a small quantity of another sublimate rises, similar to the vapour of tellurium, inasmuch as it fuses when heated; but the drops become brown, and, on cooling, yellowish and opaque; whereas those of tellurium become colourless and transparent, at least when they are in a thin layer. After a portion of the sulphur is burnt away, the assay boils up, and sputters about like the preceding sulphuret; it leaves a regulus of bismuth, which, on cupellation, tinges the bone ashes pure orange yellow.

Remark.—It appears from these experiments, that the native bismuth, found near 50 years since at Gregers-klack, near Bispberg, is properly a sulphuret of bismuth, in which the proportion of sulphur is less than in the sulphuret from Riddarhytta, which appears capable of being brought by roasting to the same degree of saturation as the preceding. The *Wasserbleysilber*, discovered by Von Born, and which, according to Klaproth's analysis, should be a sulphuret of bismuth containing only five per cent. of sulphur, is in fact quite another combination, as we shall see presently.

3. *Alloys of tellurium and bismuth.*

a. Alloy of tellurium, selenium, and bismuth, from Norway. Native Tellurium of Esmark. (I am indebted to M. L'Abbé Haüy for the specimen which was the subject of the following experiments.)

Alone, on charcoal, fuses into a metallic globule, which imparts a blue colour to the flame and exhales a strong odour of selenium. A white, pulverulent deposit, iridescent at the edges, forms on the charcoal, which disappears before the reducing flame, colouring it green. By a prolonged blast, the remaining metallic globule may be wholly dissipated. If a little salt of phosphorus be fused on the spot where the globule lay, it will indicate traces of copper.

In the open tube fuses and gives off a copious white vapour, which, after some time roasting, deposits a reddish substance in that part of the sublimate nearest the assay. This red substance is selenium, whose odour is strongly perceptible in the current of gas which issues from the tube. Heat melts the white sublimate into clear transparent drops; therefore it is oxide of tellurium. A metallic globule remains on the glass, which gives off no more vapour, and becomes covered with a fused mass of a brown colour, which, on cooling, is opaque and livid yellow; consequently, it is bismuth.

b. Alloy of bismuth and tellurium, Wasserbley-silber, of Von Born. (The assay which gave the following results, was part of a specimen from the collection of the University of Berlin, presented to me by Professor Weiss.)

Alone, in the open tube, the assay, in the form of a spangle, became brown before it fused, readily melted into a globule, and then, for a few seconds, exhaled the odour of selenium; when ignited, it gave off an abundant white vapour, which adhered to the glass, and was fusible into white transparent drops; it therefore was tellurium. What remained was a globule of bismuth, which gave no more vapour, and, by a prolonged blast, became surrounded by a fused brown oxide of bismuth, in the same manner as the pure metal.

4. *Oxide of bismuth.*

(See Bismuth, p. 150.) It sometimes gives traces of iron and copper.

8. *Tin.*

Oxide of tin.
(See the oxides of tin, p. 112.) The dark coloured species, when treated with soda, on platina foil, give more or less decisive traces of manganese.

Remark.—When columbium occurs with oxide of tin, as at Finbo, near Fahlun, its presence is ascertained by two indications; 1st, the oxide of tin reduces with more difficulty and less perfectly, leaving a pretty considerable portion not reduced; 2d, when dissolved in certain proportion with borax, the flux acquires the property of becoming opaque by flaming, or even by merely cooling.

9. *Lead.*

1. *Sulphuret of lead.*[1]
Alone on charcoal globules of lead form on the surface after the sulphur is driven off, till which

---

[1] One atom of lead, 104 + 2 atoms of sulphur, 32 = 136. C.

time the assay does not fuse; we ultimately obtain a bead of lead. By cupellation, we ascertain if it contain silver. After cupellation, the colour of the bone ashes teach us whether the lead was pure or not; in the first case, it is pure pale yellow; if it contain copper it is greenish, iron, black or brownish, &c. The roasting, as well as the cupelling, may be performed on bone ashes.

In the tube galena disengages sulphur, and gives a white sublimate of sulphate of lead, which, in a strong heat, becomes grey, even in the upper part nearest the assay. By a good flame we may fuse the sublimate, but it immediately fixes again, and gives off no volatile substance.

Remark.—The galenas of Fahlun, and the copper ores of Åtvidaberg develope the odour of selenium, when roasted on charcoal; and if the roasting be performed in a tube, a small, but very visible red sublimate of selenium may be obtained. To effect this, the roasting must be performed very gently, and carried very far, for the selenium does not begin to separate till towards the end of the operation. A red ring then forms at an inch from the assay, and the smell of selenium begins to be evident in the upper part of the tube. The selenium may be concentrated by exposing the part of the tube, between the assay and the ring formed by the sublimate, to the flame of a taper, so as to drive the portion of selenium, which has been deposited in the intermediate space, towards the ring. If the quantity of selenium be slight, the red ring is scarcely discernible when we look across the tube; but it is easily transferred to a dark ground. If the galena contain arsenic, the selenium may easily be confounded with sulphuret of arsenic.

2. *Spiesglansbleyerz,* Bournonite, Endellione, from Bleyberg.[1]

On charcoal, fuses and gives off fumes for some time, then congeals into a black globule. In a strong heat vapour of lead is disengaged and forms a circular deposit on the charcoal: a scoriaceous mass remains behind, in which the fluxes detect the presence of a considerable quantity of copper, a globule of which metal may be obtained by soda after roasting the lead.

In the tube the odour of sulphurous acid is developed and a dense white vapour, which is deposited in great measure on the lower side of the tube; this portion of the sublimate is neither volatile nor fusible, but the upper deposit is **volatile**. The first is antimonite of lead, the second oxide of antimony.

3. *Licht Weissgültigerz,* from Freyberg.[2]

[1] According to the analysis of Mr. Hatchett, this mineral consists of lead 42·62, antimony 24·23, copper 12·8, iron 1·2, sulphur 17. The specimen was from Cornwall. *Phillips.*— Berzelius estimates its composition as that of 1 atom of sulphuret of lead, 136+1 atom of sulphuret of copper (64+16) 80+1 atom of sulphuret of antimony, 93 = 309. C.

[2] *White silver ore,* containing, according to Klaproth's analysis, silver 20·40, lead 48·06, antimony 7·88, iron 2·25, sulphur 12·25, alumina 7·0, silica, 0·25. *Phillips.*—Its theo-

Decrepitates strongly, fuses easily and gives off fumes of lead.

In the open tube it behaves like the preceding. The roasted mineral gives with the fluxes the tint of nickel, and sometimes also that of cobalt. With borax we obtain a metallic globule, which, by cupellation with lead, loses considerably in bulk, and leaves a residuum of pure silver.

4. *Dunkel Weissgültigerz*, from Sala.[1]

In the open tube behaves like the foregoing; after roasting it leaves a mass of scoriæ which with borax gives the tint of iron, and a globule of lead containing a very little silver, that may be detected by cupellation.

5. *Blättererz*. (See *Gold*, p. 142.)

6. *Oxide of lead*, red and yellow. (See *Oxide of Lead*, p. 153.)

7. *Sulphate of lead*, from Anglesea.[2]

Decrepitates; on charcoal in the exterior flame, fuses into a transparent globule, which becomes milky as it solidifies. In the reducing flame it effervesces and gives a particle of lead.

---

retical composition, according to Berzelius, is, 1 atom of sulphuret of lead, 136; 1 atom of sulphuret of silver, 142; 1 atom of sulphuret of antimony 93; and 1 atom of arseniuret of nickel (26+38) 64. C.

[1] Composed of 1 atom of sulphuret of lead, and 1 of sulphuret of antimony. C.

[2] One atom of oxide of lead (104+8) 112+1 atom of sulphuric acid (16+24) 40 = 152. The weight of an atom of oxygen is 8. Berzelius considers the protoxide of lead to contain two atoms of oxygen. C.

With borax, and *salt of phosphorus,* it behaves like pure oxide of lead.

With *glass of soda and silica* it assumes the colour of liver of sulphur at the moment the glass cools.

8. *Carbonate of lead,* from Alston Moor.[1]

Behaves like pure oxide of lead, except that it decrepitates strongly and its white colour changes to yellow by heat.

a. Chloro-carbonate of lead—Horn-lead, from Matlock, Derbyshire.[2]

Alone, in the exterior flame, fuses into a transparent globule, which becomes pale yellow on cooling.

With *oxide of copper dissolved in salt of phosphorus* it presents the usual effect of muriatic acid (a blue flame surrounding the assay globule).

b. Sulphate and carbonate of lead, from Leadhills.

The specimen used in the following assay belonged to the collection of the King of France, and was given me by Count Bournon. It is described in the catalogue of the private mineralogical collection of the King of France, (Paris, 1817. pp. 343, 344), by the name of rhomboidal carbonate of lead.

Alone, on charcoal, at first it swells up a little and becomes yellow, but resumes its white colour

---

[1] Composed of 1 atom of oxide of lead, 112+1 atom of carbonic acid, (6+16) 22 = 134. An atom of carbon weighs 6. Berzelius considers it as containing 2 atoms of carbonic acid. C.

[2] Composed of 1 atom of chloride of lead, 140+1 atom of carbonate of lead, 134 = 274. C.

on cooling; it fuses into a globule, which also is white when cold. It reduces equally well into a globule of metallic lead, either with or without soda.

With glass of silica and soda, it gives the colour of liver of sulphur, precisely like the sulphate of lead.[1]

9. *Phosphate of lead*, from Freyberg.[2]

Alone, on charcoal, fuses in the exterior flame; the globule crystallizes, and after cooling has a dark colour. In the interior flame it exhales the vapour of lead, the flame assumes a bluish colour, and the globule on cooling forms crystals with broad facets, inclining to pearly whiteness. At the moment it crystallizes, a gleam of ignition may be perceived in the globule.

With borax, salt of phosphorus, and soda, it behaves like oxide of lead.

With boracic acid and iron it gives phosphuret of iron and metallic lead, which may be obtained separately when the phosphuret congeals, the lead still remaining in fusion. This lead gives no silver by cupellation.

10. *Aseniate of lead*, from Johann Georgenstadt, and from Cornwall.[3]

Alone, on charcoal, fuses with some difficulty,

---

[1] This mineral dissolves with effervescence in nitric acid, and leaves a residuum of sulphate of lead in the form of a white *powder*. B.

[2] Composed of 1 atom of oxide of lead, $112 + 1$ atom of phosphoric acid, $28 = 140$. C.

[3] Composed of 1 atom of oxide of lead, $112 + 1$ atom of arsenic acid, $62 = 174$. C.

and is then instantly reduced into numerous globules of lead, with copious disengagement of the fume and smell of arsenic. With the fluxes it behaves like oxide of lead, with this difference, that the glass of the arseniate gives off arsenical fumes.

If we take a crystal of arseniate of lead in the forceps, and fuse its anterior extremity in the outer flame, the fused part crystallizes on cooling, in the same manner as phosphate of lead. The fused portion must not touch the platina, as it would spread over and act on the metal. It flows equally on glass.

Remark.—Arseniate of lead, containing phosphate of lead, does not completely reduce; the phosphate always remains in the state of a salt, and in the form of a globule which crystallizes on cooling. Phosphate of lead containing traces of arseniate, gives metallic lead and exhales the odour of arsenic when fused in the interior flame.

11. *Molybdate of lead*, from Bleyberg.[1]

Alone, decrepitates strongly, and acquires a brown yellow colour, which flies on cooling. On charcoal it fuses and is absorbed, leaving at the surface a portion of reduced lead. By washing the absorbed part, we obtain a mixture of malleable grains of lead and metallic molybdena, the latter has a metallic lustre, but is neither malleable nor fusible.

---

[1] Composed of 1 atom of oxide of lead, $112+1$ atom of molybdic acid, $72 = 184$. According to Berzelius, it contains 2 atoms of acid. C.

With borax, in the exterior flame, fuses readily into an almost colourless glass. In the interior flame we obtain a transparent glass, which on cooling becomes all at once dark and opaque; if we flatten it between the forceps, we perceive that its colour is brownish.

With salt of phosphorus dissolves readily; a small proportion of molybdate of lead gives a green glass, as molybdic acid does, a larger quantity gives a black opaque glass.

Soda dissolves molybdate of lead, a portion of the mass is absorbed by the charcoal, and reduced lead remains on the surface.

12. *Chromate of lead*, from Siberia.[1]

Alone, decrepitates, splits in the direction of the axis of the crystal and assumes a deeper colour, which becomes clear on cooling. On charcoal it fuses and flows abroad, at the same time the lower part is reduced and developes the flame, and fumes peculiar to lead. The upper part is a mass of a dark colour, which gives a brown red powder and does not become green by heat.

With borax dissolves readily. A small quantity of the chromate colours the glass green; a larger quantity in the exterior flame gives a glass of a green colour, but so loaded with blackish particles that it appears opaque. In the

---

[1] Composed of 1 atom of oxide of lead, 112+1 atom of chromic acid, 52 = 164. C.

reducing flame its colour is dark, and on cooling it assumes the aspect of a greenish grey enamel.

With salt of phosphorus, fuses easily into a fine green glass. A larger quantity of the chromate gives a glass which is opaque, and grey or greyish green on cooling.

With soda, on charcoal, the mass is absorbed, and grains of metallic lead are produced. On platina, in the oxidating flame, it forms a liquid saline mass of a brown yellow colour, which becomes yellow when cold. In the reducing flame the fused mass is green.

13. *Tungstate of lead*, from Zinnwald.[1]

(The specimen was furnished me by Mr. Breithaupt).

Alone, on charcoal, fuses and gives off vapour of lead, leaving a crystalline globule, of a dark colour and metallic aspect, but which gives a clear grey powder.

With borax, in the exterior flame, dissolves without colour; in the interior flame before a brisk blast it becomes yellowish, and on cooling grey and opaque. By a graduated flame the lead is dissipated in fumes, and the globule on cooling is transparent and dark red, like that obtained with pure tungstic acid.

With salt of phosphorus in the exterior flame we obtain a colourless glass, and in the interior

---

[1] Composed of 1 atom of oxide of lead 112+2 atoms of tungstic acid 240 = 352.

flame a bright blue glass, whose colour is not altogether so pure as that from tungstic acid. A larger proportion of the tungstate gives a greenish glass, which in the end is opaque.

With soda, we obtain a large quantity of globules of lead.

14. *Plomb-gomme,* from Huelgoat.[1]

(I am indebted to M. Gillet de Laumont for this substance).

Alone, in the matrass, gives off aqueous vapour, the assay sometimes decrepitating violently.

On charcoal it becomes opaque, whitens, swells up like a zeolite, and by a strong heat is brought into semifusion, but cannot be perfectly fused.

With borax, fuses into a colourless transparent glass.

With salt of phosphorus, similar solution and similar result. With a certain proportion of the plomb-gomme the glass becomes opaque on cooling.

With soda no solution; but globules of lead start out on all sides.

With nitrate of cobalt we obtain a fine pure blue colour.

10. *Copper.*

1. *Sulphuret of Copper.*[2]

---

[1] Composed of 1 atom of oxide of lead $112+6$ atoms of alumina $102+6$ of water $54 = 268$. Thomson's Chemistry, vol. iii. p. 542. Berzelius considers it as a quadri-aluminate of lead, with 12 atoms of water.

[2] Composed of 1 atom of copper $64+1$ atom of sulphur $16 = 80$.

Alone, on charcoal, exhales the odour of sulphurous acid, fuses easily in the exterior flame, and projects drops in a state of ignition. In the interior flame, it becomes covered with a crust, and then is no longer fusible. This experiment may be frequently repeated. As long as any sulphur remains no copper separates, whence it seems that sulphur and copper may be melted together in any proportion.

In the open tube, sulphurous acid is evolved, and a part of the assay burns, but no sublimate is formed. The roasted mineral gives a globule of copper, with either soda or borax.

2. *Silber-kupferglanz,* (Haussman and Stromeyer). *Argentiferous sulphuret* of Copper. (Bournon), from Ecatherinenburg.[1]

Alone, fuses readily, exhales the odour of sulphurous acid, gives off no fumes (not even in the tube), does not oxidate, nor project any scoriæ. The globule has a grey colour, metallic lustre, is slightly coloured on the surface,[2] is semi-malleable, and has a grey fracture. With the fluxes it produces the effects of copper. By cupellation with lead on bone ashes, it gives a large globule of silver, and the cupel is coloured blackish green.

3. *Compound of sulphuret of antimony, and*

---

[1] Composed of 2 atoms of sulphuret of copper 160+1 atom of sulphuret of silver 142 = 302. C.

[2] *Se colore légerèment à la surface,* this seems redundant, but it is so in the original. C.

sulphuret of copper. (Graugültigerz). a. *sulphuret of copper and antimony*, from Ecatherinenburg. (Bournon, catalogue of the collection, &c. p. 235.) b. *Endellione,* or *Bournonite,* from Saint-Harey, near Grenoble. c. *Schwarzerz,* from Kapnick.

Alone, in the open tube, fuses and gives off fumes of antimony, which contain scarcely any antimonious acid; exhales the odour of sulphurous acid, not, however, very evident till after blowing a few seconds; completely bleaches brazil wood paper, placed in the upper part of the tube. The roasted mineral solidifies into a black mass.

Alone, on charcoal, deposits antimony; no trace of the vapour of lead. The globule diminishes in bulk, becomes grey and semi-malleable; with borax, it retains its grey colour for some time, and then produces the characteristic effects of copper; fused with soda it gives a globule of copper.

4. *Copper pyrites,* a compound of sulphuret of copper and sulphuret of iron.

Alone, on charcoal, on the first impulse of the flame, it assumes a superficial dark coloured tinge, and blackens; becomes red on cooling, fuses more readily than sulphuret of copper, and affords a globule, which, after the blast has been kept up some time, is attractable by the magnet; it is brittle, and has a greyish red fracture. If after long exposure to the oxidating flame, we treat it with a very small quantity of borax, it gives metallic copper.

In the open tube exhales a strong odour of sulphurous acid, but gives no sublimate.

In the matrass, no sulphur sublimes.

After roasting, this mineral produces a compound effect with the fluxes, derived from that peculiar to each of its ingredients, the iron and copper.

With soda, we obtain separate globules of iron and copper, provided the sulphur has been perfectly burnt away.

5. *Sulphuret of tin*, (Zinnkies), from Cornwall.[1]

Alone, fuses at a high temperature, and exhales, in the exterior flame, the odour of sulphurous acid; becomes snow white on the surface, and covers the charcoal with a circle of white powder, extending about a quarter of an inch from its base round the assay. This powder, which constitutes the principal pyrognostic character of the mineral, is oxide of tin. It differs from the pulverulent deposit of other volatile metals, 1st, in being contiguous to the assay; 2ndly, in not being volatile either in the exterior or interior flame.

In the open tube it exhales the odour of sulphurous acid, and is covered with fixed white fumes: a portion of which condenses also on the sides of the tube close to the assay.

After long roasting on charcoal, we obtain a grey

---

[1] *Bell metal ore.*—Composed of 1 atom of sulphuret of tin $(59+32)\ 91 + 2$ atoms of sulphuret of copper $160 = 251$. C.

brittle metallic globule, which gives with the fluxes the effects of iron and copper. With a mixture of soda and borax in the oxidating flame, sulphuret of tin gives a dark coloured, hard, and very slightly malleable globule of copper.

6. *Nadelerz*, from Ecatherinenburg.[1]

Alone, fuses, gives off fumes, and forms a white deposit on the charcoal, slightly yellowish on its interior edge; it then gives a metallic globule resembling bismuth. The fumes are reduced in the interior flame without colouring it.

In the open tube gives white vapour, partly fusible and partly volatile; the fusible part melts into clear drops, some of which turn white on cooling; the current of air which issues from the tube diffuses the odour of sulphurous acid. The globule of bismuth is surrounded by an oxide, which is black whilst liquid, but becomes transparent and greenish yellow on cooling. With the fluxes, this globule gives the effects of copper, somewhat faintly. After a strong blast, we ultimately obtain a globule of copper, which, by cupellation with lead, leaves scarcely perceptible traces of silver.

Remark.—John found from one to two per cent.

---

[1] *Needle ore.—Jameson.—Plumbo-cupriferous sulphuret of bismuth*, Hauy.—Composed of bismuth 43·2, lead 24·3, copper 12·1, nickel 1·5, tellurium 1·3, sulphur 11·5. (*John's Analysis.*) Its theoretical composition, according to Berzelius, is 1 atom sulphuret of lead $136 + 2$ atoms sulphuret of copper $160 + 2$ atoms sulphuret of bismuth $206 = 502$. C.

of tellurium in this mineral. Its presence accords well with the properties of the vapour formed in the tube whilst roasting; but its quantity supposes a larger proportion of tellurium than John obtained.

The vapour of tellurium usually colours the reducing flame green; but this phenomenon does not occur with the needle ore, and if sometimes a slight colour appear, its tint is bluish. The same thing happens in the assay of the Blättererz, in which tellurium and lead occur together. (See Blätterez, p. 143). Thus lead modifies to a certain extent, the characteristic effects of tellurium.

7. *Seleniuret of copper*, from Skrickerum.[1]

Alone on charcoal, fuses into a slightly malleable grey globule, exhaling at the same time a very strong odour of selenium.

In the tube, gives at the same time a red pulverulent sublimate of selenium, and selenic acid, which forms, beyond the condensed selenium, crystals volatile at a very low heat.

After very long roasting, during the whole of which the assay developes the odour of selenium, we obtain by soda a globule of copper.

8. *Euchairite*, from Skrickerum.[2]

Alone fuses, exhales a strong odour of selenium, and gives a soft, but not malleable grey metallic

---

[1] Composed of 1 atom of copper 64+1 atom of selenium 41=105. C.

[2] Composed of 2 atoms of seleniuret of copper 210+1 atom of bi-seleniuret of silver (110+82) 192 = 402. C.

globule; cupelled with lead it leaves a globule of silver; the odour of selenium is perceptible during the whole operation.

In the open tube, it behaves like seleniuret of copper.

With the fluxes, the effects of copper are very decidedly produced.

9. *The Fahlerz ores.* These minerals behave differently under the same circumstances, and divide into two classes, one composed of such as give off arsenic, the other of those that give off antimony in roasting. Some kinds fuse, boil up and fume all at once; others fuse first completely and then intumesce, forming excrescences, which present in miniature a cauliflower appearance, but fuse in a strong heat. When treated with soda after previous roasting, they all give a globule of metallic copper. With the fluxes, the Fahlerz ores produce the effects due to iron and copper.

10. *Protoxide,* and

11. *Oxide of copper,* (see page 115).

12. *Neutral sulphate of copper,* as well as the *sub-sulphate,* is discoloured by heat, and gives off vapour of water. The neutral salt becomes white, and the sub-salt black. Sulphuric acid is detected in both, by pulverising the roasted assay, mixing it with charcoal powder, and heating it in a tube closed at one end; a large quantity of sulphurous acid is given off, and is evident both by its smell and by its action on moistened brazil wood paper, placed in the tube. This effect is very distinct,

even though the particle of sub-sulphate be not larger than a pin's head.

13. *Sub-muriate of copper,*[1] arenaceous and compact, from Chili.

Alone, it strongly colours the flame blue, and its edges green. A red pulverulent deposit forms on the charcoal, round the assay; this deposit colours the flame blue on the part which plays over the surface of the support. The assay fuses, and is reduced into a globule of copper surrounded with scoriæ. The arenaceous mineral gives more scoriæ than the compact. The scoriæ produce the effects derived from copper and iron; that of the latter is distinctly seen on the glass where the reduction took place, before the action of the copper is developed by cooling.

With the fluxes, the muriate behaves like oxide of copper.

14. *Phosphate of copper*, from Ehrenbreitstein.[2]

Alone, gives no colour to the flame; falls to powder in a strong sudden heat, but preserves its cohesion if the heat be gradually raised; blackens and fuses, retaining its black colour; in the centre of the mass, a small globule of metallic copper is perceptible. This nucleus of copper emits a brilliant light, or *fulguration*, at the instant it con-

---

[1] Composed of 2 atoms of peroxide of copper 160 + 1 atom of muriatic acid 37 + 4 atoms of water 36 = 233. C.

[2] Composed of 1 atom of sub-phosphate of copper (2 oxide 160 + 1 of acid, 28) 188 + 10 atoms of water 90 = 278 C.

geals, very like the *brightening* of gold or silver on the cupel.

With salt of phosphorus and borax it behaves like pure oxide of copper.

With soda, a peculiar phenomenon presents itself. A small quantity of soda gives a liquid globule. If we add a fresh quantity of soda, the mass swells up for a moment, then liquefies again, and at every fresh portion that we add to the assay, the same phenomenon occurs, till at last it dilates, and becomes solid and infusible. With a great quantity of soda, the saline mass is absorbed by the charcoal, and leaves copper on the surface.

The principal characteristic effect of phosphate of copper is that produced by fusing it with a nearly equal bulk of metallic lead. If we expose the mixture to a very good reducing flame, the whole of the copper separates in the metallic state, and a mass of fused phosphate of lead, which crystallizes on cooling, forms round the regulus. If, after the phosphate has congealed, we take off the comparatively more fusible plumbiferous metal, and then fuse the mass again, we obtain a more sperical globule, with broader crystalline facets.

15. *Carbonate of copper*, green and blue.[1]

Alone, in the matrass, gives off water, and blackens.

---

[1] The first composed of 1 atom of carbonate of copper $(80+22)$ $102+1$ atom of water $9 = 111$; the second of 1 atom of hydrate of copper $(80+18) 98 + 2$ atoms of bicarbonate of copper $(\overline{80+44} \times 2)$ $248 = 346$. C.

On charcoal, fuses and behaves in every respect like pure oxide of copper. (p. 115).

16. *Arseniate of copper* of different species, from Cornwall. Arseniate of copper behaves with the fluxes, in general, like oxide of copper, but exhales by heat a strong odour of arsenic, and when reduced alone with soda, gives a white brittle metallic globule.

(*a*). The greyish white arseniate, crystallized in capillary needles, when heated alone in the matrass gives off no water, nor experiences any change. On charcoal it is reduced, at the moment it fuses, with detonation, which causes it to penetrate deep into the support, from which we obtain a metallic globule which becomes red on cooling. The red colour is owing to a thin covering of protoxide of copper; the globule is white in the interior, and breaks under the hammer.

(*b*). The dark green crystallized arseniate behaves nearly in the same manner, but throws out a fused scoria round the reduced metallic globule, as may be seen on attentive examination. If we add lead, pour off the fused metal after the scoria has congealed (which naturally adheres to the surface of the support), and fuse the scoria again, we then obtain phosphate of lead, of a white colour, which crystallizes at the moment of congelation.

(*c*). The compact green variety, rather *blebby* internally, presents the same phenomena, but with lead gives a larger quantity of phosphate of lead.

(*d*). The beautiful variety in clear blue crystals, gives off much water in the matrass, fuses imperfectly, and is not reduced with detonation; but leaves a mass of scoriæ, amongst which may be perceived some white metallic globules. It gives no indication of iron with the fluxes.

Remark.—This last variety, therefore, contains, beside oxide of copper, some other base which is not reducible.

17. *Vauqueline*, from Siberia.[1]

Alone, in the matrass, it gives off no water. On charcoal it intumesces slightly, then fuses and froths up abundantly, and is converted into a dark grey globule, possessed of metallic brilliancy, and surrounded with small beads of reduced lead. The greatest part of the globule suffers no change, even in a very powerful reducing flame.

With borax it fuses in small quantity, with effervescence, into a green glass, which in the exterior flame retains its transparency on cooling; but after exposure to a good reducing flame, on cooling it becomes red and diaphanous, or red and opaque, or, lastly, perfectly black, according to the quantity of the mineral in the glass. A little tin added to the assay facilitates the developement of the red colour, which arises from the copper. If

---

[1] Composed of 2 atoms of sub-sesqui-chromate of copper $(\overline{80 \times 3 + 52 \times 2})$ $344 + 2$ atoms of bi-chromate of lead $(\overline{112 + 104 \times 2})$ $432 = 776$. The number of atoms is doubled to get rid of the anomalous half atom of copper in the sub-sesqui-chromate. C.

we add a large portion of the mineral to the assay all at once, the glass blackens immediately.

With salt of phosphorus, the same phenomena.

Soda dissolves it with effervescence. On the platina wire in the oxidating flame, we obtain a green transparent glass, which becomes yellow and opaque on cooling. If we put a drop of water on the glass, the alcaline chromate which is present gives it a yellow colour. The mass is absorbed by charcoal, and we obtain globules of lead by washing, &c.

18. *Kieselmalachite*, from Siberia.[1]

Alone, in the matrass, gives off water and blackens.

On charcoal it becomes black in the exterior flame, and red in the interior, but does not fuse.

With borax, readily fuses into a glass, which presents the effects of copper. Heated gently in the exterior flame, it colours it momentarily of a fine green; continuing the blast, no colour re-appears; but let the globule cool, then heat it again to redness, and the flame is coloured anew. This experiment may be repeated on the same globule at pleasure. The phenomenon does not take place with pure oxide of copper. With a good reducing flame, we obtain a globule of metallic copper in

---

[1] *Chrysocolla*, composed of 2 atoms of sub-sesqui-silicate of copper ($80 \times 3 + 16 \times 2$) $272 + 12$ of water $108 = 380$. The atoms are doubled as in the preceding. C.

the borax; the glass may be deprived of colour by the blast.

With salt of phosphorus dissolves and presents the effects of copper, but gives a sort of skeleton (squelette) of silica, distinctly perceptible on suffering the glass to cool, after having exposed it to the exterior flame. The flame is not at all coloured when we use this flux.

With soda, on charcoal, fuses into a dark coloured opaque glass, which internally is red after cooling, and contains a globule of copper. With a large quantity of soda the glass penetrates the charcoal, and leaves metallic copper on the surface.

19. *Dioptase*, from the country of the Kirguise.[1]

Behaves in every respect like the preceding mineral. The only remarkable difference is, that it takes a larger dose of soda before it is absorbed by the charcoal, and that the copper may be sufficiently reduced to render the glass perfectly colourless.

Remark.—When we treat Kieselmalachite, or Dioptase, with the fluxes, without having previously heated them red hot, they dissolve with effervescence, from the escape of water, and sometimes also from the disengagement of carbonic acid, which is not unfrequently found in Kieselmalachite.

---

[1] Tartars who inhabit the country adjoining the Caspian Sea, on the North East. C.

11. *Nickel.*

1. *Sulphuret of nickel.* (Haarkies).[1]

In the open tube, exhales the odour of sulphurous acid, bleaches brazil wood paper, and becomes black, but without changing its form.

On charcoal, in a good heat, semifuses into an agglutinated mass, which is metallic, malleable and magnetic, and consists wholly of nickel.

After roasting in contact with the air, it behaves with the fluxes like oxide of nickel.

With the glass of silica and soda, before it is roasted it develops the odour of a sulphuret.

2. *Arsenical nickel,* from Freyberg.[2]

In the matrass nothing volatile is given off; semifuses at the temparature at which glass softens, and a deposit of white arsenic forms on the sides of the glass, at the expense of the air in the matrass.

On charcoal fuses into a white metallic globule, with the disengagement of arsenical fumes and odour.

In the open tube the roasting is easily effected; a large quantity of white arsenic is formed, and a yellowish green residuum left, which by fresh roasting on charcoal, and fusion with soda and a

---

[1] Composed of 1 atom of nickel 26 + 2 atoms of sulphur 32 = 58. C.

[2] Composed of 1 atom of nickel 26 + 1 atom of arsenic 38 = 64. C.

little borax, gives a metallic globule tolerably malleable and very magnetic.

After roasting, it behaves with the fluxes like oxide of nickel, and commonly after reduction colours the glass blue, denoting thereby the presence of some oxide of cobalt.

3. *White ore of nickel*, from Loos (Nickelglanz).[1]

Alone, in the matrass, decrepitates strongly, and gives, at a red heat, a great quantity of sulphuret of arsenic, that sublimes in the form of a fused, transsparent, reddish mass, which does not become opaque on cooling. The residuum has the appearance of arsenical nickel, and behaves like that mineral with the fluxes. The effects indicative of iron are difficult to produce; commonly the blue colour of cobalt only is perceptible after the nickel is extracted by reduction.

4. *Nickel-spiesglanserz*, from Baudenberg, in the Comtè de Nassau. (The mineral used in the following experiment was furnished by Professor Ullmann).[2]

In the open tube gives off abundant fumes of antimony and a slight odour of sulphurous acid;

---

[1] Composed of 1 atom of bisulphuret of nickel (26+64) 90+1 atom of biarseniuret of nickel (26+76) 102 = 192. C.

[2] *Antimoine sulphuré nickelifere.* Hauy. According to John, it contains antimony with arsenic 61·68, nickel 23·33, sulphur 14·16, silica with silver and lead 0·83, trace of iron. Jameson. C.

it bleaches brazil wood paper placed in the tube.

In the matrass a small portion of white sublimate is formed, apparently at the expense of the air in the vessel.

On charcoal fuses and fumes considerably; a slight arsenical odour is sometimes perceptible, but not easily distinguished. However far we push the roasting of the metallic globule, it always remains fusible and brittle. With soda it neither gives off the odour nor vapour of arsenic; the soda is not absorbed by the charcoal, but remains on the surface, where it fuses with the mineral into a black globule. The glass formed with this flux displays the colour of a sulphuret. With the other fluxes the metallic globule produces only the characteristic effects of cobalt.

5. *Arseniate of nickel*, (pulverulent, slightly greenish white) from Allemont.[1]

Alone, in the matrass, gives off water, and its colour becomes darker.

On charcoal exhales a strong odour of arsenic; in the interior flame fuses into a globule of arseniferous nickel.

With the fluxes it behaves like oxide of nickel,

---

[1] Composed of 2 atoms of sub-sesqui-arseniate of nickel $(\overline{26 \times 3} + \overline{62 \times 2})$ 202 + 18 atoms of water 162 = 364. The atoms doubled as before. C.

Berthier Annales de Chimie et de Physique, tome xiii. page 57. B.

but in the reducing flame betrays the presence of a pretty considerable quantity of cobalt.

6. *Pimélite*, from Kosemütz.[1]

Alone, in the matrass, blackens and gives off water, which smells of petroleum. The black colour is owing to a portion of charcoal contained in the mineral; when this is burnt away, the mass is greenish grey, inclining here and there to brown. Pimelite is infusible, but becomes converted into scoriæ in the thin parts, and assumes a dark grey colour.

With borax dissolves and displays the effects of nickel; after the reduction of which, it gives no trace of cobalt.

With salt of phosphorus fuses, in small quantity, into a transparent glass; with a larger portion of the assay we obtain a glass, whose colour indicates the presence of nickel, and which holds in suspension a white skeleton of undissolved silica.

With soda fuses imperfectly into a nearly globular scoriaceous mass. A larger portion of soda causes it to be absorbed by the charcoal, from which we obtain a very large quantity of reduced nickel by washing.

Remark.—Pimélite, externally, has considerable resemblance to talc supposing it to contain nickel. Its property of blackening, and of exhaling an empyreumatic odour when heated in a close vessel, is

---

[1] Consists of oxide of nickel 15·62, silex 35, alumine 5·10, lime 0·40, magnesia 1·25, water 37·91. Klaproth. (*Phillips.*)

also common to almost all the minerals which contain magnesia as an essential constituent, that is in notable quantity. Hence its composition requires a more particular examination. Possibly oxide of nickel may have in this case the same property as magnesia, which it resembles also in forming double salts with potassa and ammonia.

12. *Cobalt.*

1. *Sulphuret of Cobalt,* from Bastnäs, near Riddarhytta.[1]

Alone, in the matrass, gives off nothing volatile, though, from its composition, a sublimate might have been expected; does not decrepitate.

In the open tube gives sulphurous acid, and a trifling quantity of concentrated sulphuric acid, which sublimes in the form of little drops, distinguishable by the microscope and of a white colour. If the tube be dusty they become black, from their action on the dust. The sulphuric acid is produced at the beginning of the roasting, and its quantity does not increase as the operation proceeds. No trace of arsenic can be discovered in the assay.

On charcoal fuses after roasting into a grey metallic globule, from which it is difficult to drive off the last portions of sulphur.

[1] *Cobalt Pyrites.* Constituent parts, according to Hisinger's analysis, cobalt 43·2, sulphur 38·5, copper 14·4, iron 3·53, (Jameson). Berzelius considers it composed of 1 atom of bisulphuret of iron + 4 atoms of sulphuret of copper + 12 atoms of sulphuret of cobalt. C.

With the fluxes the effects of the cobalt predominate so much that it is impossible to distinguish those of the iron and copper. But if we repeatedly remelt the grey globule with borax in the exterior flame, (it having been previously fused on charcoal) the borax seizes on the cobalt, and the copper collects together (se concentre); so that, when we afterwards fuse the mass with salt of phosphorus, and expose the glass, saturated with metallic matter, to the reducing flame, the red colour of the protoxide of copper is developed on cooling, although tinged blue by the oxide of cobalt.

2. *Arsenical cobalt*, from Riegelsdorf.[1]

Speisskobalt, from Freyberg and Bieber.

Alone, in the open tube, very readily gives off arsenious acid.

In the matrass some species give a little metallic arsenic—others none at all.

On charcoal they all exhale the vapour and odour of arsenic, and fuse into a white metallic globule, which remains brittle even after long treatment with borax, which it colours cobalt blue.

3. *Scapiform arsenic*, (Stänglicher arsenik) from Schneeberg.

Scherbenkobalt, from Saxony.

---

[1] Arsenic 74·22, cobalt 20·31, iron 3·42, sulphur 0·89, copper 0·16; or, arseniuret of cobalt 51·70, arseniuret of iron 9·17, persulphuret of iron 1·55, sulphuret of copper 0·20, arsenic 36·38. *Stromeyer.* An. de Chim. viii. 81. Berzelius considers it to consist of 1 atom of bi-arseniuret of cobalt $(26+\overline{38\times 2})$ 102, and arsenic. C.

In the matrass these minerals give off no arsenic, or only an exceedingly small quantity.

In the open tube we obtain an abundant sublimate, partly consisting of arsenious acid, and partly of metallic arsenic, and a grey brown infusible residuum, which, after a thorough roasting, readily dissolves in the fluxes.

With borax they combine and give the blue colour indicative of cobalt, without any other effect.

With salt of phosphorus both ores fuse alike, and exhibit the colour of cobalt; but the *scapiform arsenic* very decidedly gives traces of nickel, for it does not assume the blue colour till the nickel has been precipitated by tin.

4. *Koboltglanz*, from Tunaberg.[1]

Alone, in the matrass, not the least change.

In the open tube roasts with difficulty; does not give off any arsenious acid without a strong heat; exhales the odour of sulphurous acid, and whitens brazil wood paper placed in the tube.

On charcoal fumes abundantly, and, after roasting some time, fuses, when it behaves like fused arsenical cobalt.

5. *Black oxide of cobalt*, Schwartzer Erdkobalt. (Its locality not known.)

Alone, gives off empyreumatic water.

---

[1] Cobalt 44, arsenic 55, sulphur 0·5. (*Klaproth.*) Jameson. Berzelius thinks it may be composed (adding a ? to his formula) of 1 atom of bi-arseniuret of cobalt + 1 atom of bisulphuret of cobalt. C.

On charcoal exhales a slight arsenical odour; does not fuse.

With borax and salt of phosphorus dissolves, and gives so deep a tint of cobalt blue, that no other effect is discernible.

With soda it is infusible. On the platina wire it gives a mass coloured deep green by manganese. If we separate the green portion, and heat it alone on charcoal, we obtain a scarcely magnetic, white metal, which communicates an iron tinge to salt of phosphorus, and also the property of turning milk white on cooling. I have not more particularly examined this metal.

6. *Arseniate of cobalt*, (crystallized) from Schneeberg.[1]

Alone, in the matrass, gives off water and turns brown, but no sublimate rises.

On charcoal fumes abundantly, and exhales the smell of arsenic. Fuses in a good reducing flame, and is converted into arsenical cobalt.

With the fluxes we obtain a blue glass.

7. *Arsenite of cobalt* (pulverulent), from Schneeberg.

Alone, both *in the matrass* and *open tube*, gives

---

[1] Composed of 1 atom of sub-sesqui-arseniate of cobalt $(\overline{34 \times 3} + \overline{62 \times 2})\ 226 + 12$ atoms of water $108 = 334$. According to the formula in the Tables, Nouveau Système, p. 204, and in the Essai sur la Theorie des Proportions, &c. The formula in the present work gives 4 atoms of oxide of cobalt and 2 atoms of arsenic acid, which denotes a sub-arseniate. C.

a large quantity of arsenious acid, after which it behaves like the preceding.

13. *Uranium.*

1. *Protoxide of uranium,*[1] from Johan Georgenstadt.

Alone, neither fuses nor alters, but in the forceps colours the external flame green.

With borax and *salt of phosphorus* behaves like oxide of uranium. (See p. 99.)

With soda does not dissolve; but when we attempt to reduce it, it gives white metallic globules of iron and lead.

2. *Yellow hydrated oxide of uranium,* from the same place.[2]

a. Pulverulent oxide of uranium, in the form of a lemon-yellow powder.

In the matrass gives off water and assumes a red colour, which it retains whilst hot; in the reducing flame it becomes green without fusing; in other respects it behaves like pure oxide of uranium.

b. Compact oxide of uranium.

Alone, in the matrass, gives off water and becomes reddish. In a strong heat, on charcoal, it fuses into a black globule.

With the fluxes behaves like the preceding.

---

[1] One atom of uranium $125 + 2$ atoms of oxygen $16 = 141$. C.

[2] One atom of peroxide of uranium $149 + 10$ atoms of water $90 = 239$. C.

With soda gives off vapour of lead, and affords white metallic globules.

Remark.—The last variety is a mechanical mixture of hydrate of uranium, uranate of lime, and uranate of lead.

14. *Zinc.*

1. *Zincblende.*[1]

Alone, sometimes decrepitates strongly; experiences little change at a red heat. Does not fuse, but the thinner parts on the edges are a little rounded by the most intense heat we can produce. Exhales but a very slight odour of sulphurous acid, and is difficult to roast.

In the open tube gives off no fumes, and suffers little change.

On charcoal it forms an annular deposit of the vapour of zinc when strongly heated in the exterior flame.

Soda attacks it feebly; but the zinc is reduced and, in a good flame, burns, and flowers of zinc are deposited on the charcoal.

2. *Oxide of zinc,* from America.[2]

Alone, suffers no change, except that its colour becomes obscure whilst hot. In the reducing flame it covers the charcoal with zinc fumes.

With borax dissolves readily, and, in the exterior

---

[1] *Blende.* One atom of zinc 34 + 1 atom of sulphur 16 = 50. C.

[2] Oxide of zinc, with oxide of manganese. C.

flame, the glass is coloured by manganese: when saturated, it becomes opaque on cooling, as well as by *flaming*.

With salt of phosphorus fuses readily into a colourless glass. The colour indicative of manganese does not appear till the glass is so saturated that it becomes opaque on cooling.

With soda no solution. No effect when we attempt to reduce it.

On platina foil a pale green colour.

With solution of cobalt the pulverised mineral (whose colour is yellow) becomes tinged greenish yellow on the edges, but gives no shade of blue, nor experiences the slightest fusion.

3. *Sulphate of zinc,* from Fahlun.[1]

Alone, in the matrass, gives off water. Heated with charcoal powder, it disengages large quantities of sulphurous acid. It gives the colour of liver of sulphur to the glass of silica and soda, when fused with it.

It behaves with the fluxes like oxide of zinc, producing, besides, the effects derived from iron.

4. *Carbonate of zinc* (crystallized) *calamine.*[2]

Alone, gives off no water; but, by heat, becomes like white enamel, and then behaves as pure oxide of zinc.

Remark.—If the calamine contain cadmium,

---

[1] One atom of oxide of zinc $42+1$ atom of sulphuric acid $40+7$ atoms of water $63 = 145$. C.

[2] One atom of oxide of zinc $42+1$ atom of carbonic acid $22 = 64$. C.

when exposed on charcoal to the reducing flame, on the first impulse of the heat, a red or orange coloured ring surrounds the assay. This appearance is very distinct when the charcoal is quite cold. (See oxide of cadmium, p. 104.)

5. *Sub-carbonate of zinc* (earthy), *Zinkblüthe,* from Bleyberg and the East Indies.[1]

Alone, in the matrass, gives off aqueous vapour. In all other respects behaves like oxide of zinc. By a continued blast it may be volatilized in the reducing flame, when it leaves a minute portion of scoriæ, in which the fluxes detect the presence of iron.

6. *Double carbonate of zinc and copper,* from Siberia.

Alone, in the matrass, it gives off water, and its green colour changes to black. With the fluxes it exhibits the effects of copper, but the globules become opaque on cooling, in consequence of the oxide of zinc which they contain.

With soda, in the experiment of reduction, we obtain a globule of copper, and the charcoal becomes covered with zinc fumes.

7. *Silicate of zinc.* Zinkglas.[2]

Alone, in the matrass, decrepitates slightly, gives off water, and becomes milk white. Intumesces a little in a strong heat, but does not fuse.

---

[1] *Earthy calamine.* Oxide of zinc 71·4, carbonic acid 13·5, water 15·1. Smithson. Phil. Trans. C.

[2] *Electric calamine.* A specimen from Rezbanya, in Hungary, analysed by Smithson, gave oxide of zinc 68·3, silica 25, water 4·4. *Jameson.*

With borax fuses into a colourless glass, which does not become milky either by flaming or cooling.

With salt of phosphorus fuses into a colourless glass that becomes opaque on cooling. By the addition of a large quantity of the assay, and not otherwise, some indications of silica are perceptible in the globule.

With soda it does not fuse, but swells up, and scarcely gives off any fumes of zinc.

With solution of cobalt becomes green at a moderate temperature; in a strong heat, the edges of the assay are coloured of a fine clear blue, whose tint is very beautiful; at the same time, we perceive an incipient fusion, and the blue colour extends a little towards the unfused portion.

8. *Gahnite*, from Fahlun.[1]

Alone, unalterable.

With borax, and *salt of phosphorus*, fuses, but with such difficulty, that these fluxes might be supposed to have no action on it. Even in powder it fuses in very small quantity.

With soda, does not fuse, but agglutinates into a dark coloured scoria.

Reduced to fine powder, and intimately mixed with soda, it gives, in the reducing flame, a very evident areola of the fumes of zinc, which surrounds

---

[1] *Automalite—Zinciferous spinelle.* Vauquelin's analysis gave, alumina 42, oxide of zinc 28, silica 4, oxide of iron 5, sulphur 17. *Phillips.* Berzelius considers it, from Ekeberg's analysis, as a quadri-aluminate of zinc, containing 1 atom of oxide of zinc $42 + 4$ atoms of alumina $68 = 110$. C.

the assay, at the beginning of the blast. This, in the order of its pyrognostic characters, is the principal distinguishing mark of gahnite.

With a mixture of soda and borax it fuses into a transparent glass, coloured by iron.

15. *Iron.*

1. *Native iron* of the meteoric stones. The blowpipe is incapable of detecting the nickel combined with the iron in these substances. I have not had an opportunity of trying the fossile iron from Kamsdorf.

2. *Sulphuret of iron.*

 a. *Magnetical Pyrites*, from Utö.[1]

Alone, in the matrass, no change.

In the open tube, gives off sulphurous acid, without any trace of sublimate.

On charcoal, in the exterior flame, becomes red, and is converted by roasting into oxide of iron. In the interior flame, it fuses, at a pretty high temperature, into a globule, which continues red hot a few seconds after it is withdrawn from the flame. After cooling, it is found covered with an unequal, crystalline black mass. Its fracture is crystalline, with a yellowish colour and metallic brilliancy.

 b. *Common sulphureous pyrites.*[2]

---

[1] Composed, according to Hatchett; of iron 63·5, sulphur 36·5. *Phillips.* Berzelius calculates it to be a compound of 1 atom of bisulphuret of iron $(28+32)$ $60+6$ atoms of sulphuret of iron $(28+16 \times 6)$ $264 = 324$. C.

[2] A bisulphuret of iron. C.

Alone, in the matrass, exhales the odour of sulphuretted hydrogen, and gives off sulphur. Towards the end of the roasting, we obtain, by urging the heat, a reddish sublimate, less volatile than sulphur, and whose quantity is variable with different pyrites; it has every appearance of sulphuret of arsenic. Well roasted sulphureous pyrites has a metallic aspect, is porous, attractable by the magnet, and behaves like magnetical pyrites.

On charcoal, its habits are similar to those of the magnetic pyrites.

3. *Mispickel* (crystallized, from various places).[1]

Alone, in the matrass, first gives a sublimate of red sulphuret of arsenic, then a black sublimate; lastly, in a strong heat, a grey crystalline sublimate of metallic arsenic rises. The residuum, when heated on charcoal, gives no more arsenical odour, and behaves like magnetic pyrites.

On charcoal mispickel at first gives off dense arsenical vapours, then fuses, exhaling the odour of arsenic, into a globule having the appearance of magnetic pyrites. If the mispickel contain cobalt, we detect it after it has been thoroughly roasted and fused with borax or salt of phosphorus in the reducing flame, by the glass assuming, when cold, its blue characteristic colour.

---

[1] *Common arsenical pyrites.* Chevreul's analysis gave, arsenic 43·4, iron 34·9, sulphur 20·1 (*Jameson*); from which Berzelius calculates its composition as 1 atom of bi-arseniuret of iron $(28 + \overline{38 \times 2})$ $104 + 1$ atom of bisulphuret of iron $(28 + 32)$ $60 = 164$. C.

4. *Graphite.*[1]

After long exposure to the interior flame, it becomes yellow or brown; is infusible, and not acted on by the fluxes.

5. *Columbiferous iron*, from Kimito, in Finland; *Cinnamon powder tantalite*, of Ekeberg.[2]

Alone, no change.

With borax its solution is extremely difficult; it is only by reducing the mineral to very fine powder that it can be made to dissolve in this flux, and even then the solution is effected very slowly. The glass has a bottle green colour, and until the whole powder is dissolved, (which requires several minutes) it gives no indication of iron in the exterior flame, and does not become opaque by *flaming* after it is cold. When the whole of the assay is dissolved, the glass behaves like that of columbite. (See further on.)

The difference in the phenomena presented by columbiferous iron and columbite, arises from a large portion of the iron and columbium in the former being in the metallic state, and insoluble without previous oxidation.

With salt of phosphorus the solution is much less difficult than with borax, and the assay is similar to that of columbite, containing no tungsten.

With soda, on platina, the action of manganese is evident. No solution.

[1] *Plumbago*, or *black lead;* according to Berthollet's analysis it consists of carbon 90·9, iron 9·1. C.

[2] *See* Afh. i Fysik, &c. B. vi. p. 237. B.

6. *Magnetic iron ore*, and
7. *Peroxide of iron*.[1]

They behave as already described under the head oxides of iron, p. 105. They are sometimes found mixed with a small quantity of chrome or titanium, which are easily detected by the effects to be described when we treat of chromiferous and titaniferous iron.

8. *Sulphate of iron*.
 a. *Iron vitriol*.
 b. *Red vitriol* (from a shaft in the mine at Fahlun, called Insjö).
 c. *Ochre vitriol*.

Heated in the matrass they all give off water, and, at a red heat, sulphurous acid, which is known by its smell and its action on moistened brazil wood paper.

After being heated red, the sulphate behaves with the fluxes like pure oxide of iron. If we treat this oxide with soda before the sulphuric acid is completely driven off, we obtain in the reducing experiment yellow metallic grains of magnetic pyrites. If the quantity of sulphuric acid combined with the oxide of iron, be so small that its presence cannot be immediately ascertained by the odour or action of the sulphurous acid disengaged from the ignited assay, it may be rendered evident, by

---

[1] The analysis of the first by Berzelius gave peroxide of iron 71·86, protoxide of iron 28·14. He considers it as composed of 1 atom of the latter, and 2 atoms of the peroxide. No. 7 is merely (according to the formula) peroxide of iron. C.

treating the oxide of iron with soda on platina foil, and dissolving a little glass in the portion of the mass dissolved by the soda. However small a quantity of sulphuric acid the assay may contain, the liver colour of the sulphuret will be developed.

9. *Phosphate of iron* (in bluish transparent crystals), from St. Agnes, in Cornwall.

Alone, in the matrass, gives a great deal of water, intumesces, and becomes sprinkled with grey and red spots.

On charcoal, intumesces, reddens by the heat, and then very readily fuses into a steel coloured globule, with a metallic lustre.

With borax, and salt of phosphorus, behaves like oxide of iron.

With soda on charcoal, in the reducing flame, gives grains of iron, which are attractable by the magnet. On platina foil there is no indication of manganese.

With boracic acid dissolves readily, and by the addition of metallic iron, in the manner detailed at p. 129, gives a fused regulus of phosphuret of iron.

All the varieties of phosphate of iron that I have had an opportunity of examining, behave in the same manner.

10. *Carbonate of iron.*

In the matrass gives off no water. Some species decrepitate violently. In a very gentle heat it blackens, and gives protoxide of iron, very attractable by the magnet.

11. *Arseniate of iron.*

(*a*). *Scorodite*, from Graul, near Schwartzenberg.

Alone, in the matrass, it first gives off water, and becomes grey white, or yellowish. In an intense heat white arsenic sublimes in small brilliant crystals, and the mass blackens. When cold, it has spots here and there on its surface, some red, others dark green; when pounded, it gives a clear grey yellow powder.

On charcoal exhales abundant fumes of arsenic, and, in the reducing flame, fuses into a grey scoria, with metallic brilliancy, and attractable by the magnet. When dissolved in the fluxes, the scoria exhibits the effects of iron, and the glass exhales a strong smell of arsenic.

(*b*). *Würfelerz*, from Cornwall.[1]

Alone, in the matrass, gives off water, and turns red. Affords but little or no white arsenic in a strong heat, intumesces slightly, becomes red on cooling, and, when pounded, gives a red powder.

On charcoal it behaves with the fluxes like the preceding mineral.

(*c*). *Eisensinter* (Eisenpecherz of Klaproth), from Freyberg.

Alone, in the matrass, gives off a very large quantity of water, and also, at an incipient red heat, sulphurous acid, easily recognised by its

---

[1] *Cube-ore.* Chenevix's analysis gave—arsenic acid 31, oxide of iron 45·5, oxide of copper 9, silica 4, water 10·5. *Jameson.* The formula represents it as composed of 1 atom of arseniate of protoxide of iron, 2 atoms of arseniate of peroxide of iron, and 10 atoms of water. C.

odour, and its action on brazil wood paper. No sublimate.

On charcoal contracts, or *shrinks* (*se retire*), gives off a thick white vapour, and a strong odour of arsenic is, for a long time, perceptible. In other respects it behaves quite like the preceding minerals.

In salt of phosphorus dissolves, and then, being exposed to the reducing flame till the whole of the arsenic is dissipated in fumes, it gives, with a small quantity of tin, a red globule, whose colour is derived from protoxide of copper. If we add the tin before the total disengagement of the arsenic, the globule turns black on cooling, and the appearance of copper cannot be produced.

Remark.—These experiments prove that Klaproth's analysis of Eisenpecherz (Beytrage, v. 221) is inaccurate, and that this mineral is a mixture of sulphate and arseniate of iron, in proportions at present unknown.[1]

12. *Chromiferous iron*, from various places.

Alone, experiences no change, except that those species, which, before they have been exposed to the reducing flame are not attracted by the magnet, are rendered obedient to it by the action of that flame.

With borax, and salt of phosphorus, solution slow, but complete. The characteristic colour of iron (with the modifications it experiences in passing from the exterior to the interior flame) is only

[1] Klaproth's analysis gave—oxide of iron 67, sulphuric acid 8, water 25. *Phillips.* C.

apparent whilst the assay is hot; but when, on cooling, its colour disappears, the fine green of the oxide of chrome is developed. This effect is much more intense if the assay has been acted on by the reducing flame, and appears in full brilliancy by the addition of tin.

Soda, on platina foil, does not attack this mineral, nor receive any colour from it when they are fused together. The reducing experiment on charcoal gives iron.

13. *Titaniferous iron* (Chrichtonite, Menachanite, Nigrine, Iserine, volcanic iron, iron sand, and, in general, every magnetic iron ore with a vitreous fracture).

Alone, infusible, unalterable. With the fluxes behaves like pure protoxide of iron; but, if we dissolve it in salt of phosphorus, and effect the complete reduction of the glass, after the colour from the oxide of iron has disappeared it assumes a red colour, of various intensity, but deepest at the last instant of cooling. The proportion of titanium is indicated by the intensity of the colour. If the protoxide of titanium be present in considerable quantity, its characteristic effect may be produced by tin; otherwise, the colour vanishes after fusing the mineral with that metal. See what has been said already on the reactions of the oxide of titanium, p. 94.

Remark.—When we submit oxide of iron, not containing titanium, to the preceding trials, the glass in which it is reduced presents, after cooling, a yellowish or reddish tint, which an inexperienced operator, expecting to find a perfectly colourless glass, might mistake for an indication of a small quantity of titanium. In general, the presence of that metal is not to be suspected, if the effect be equivocal.

14. *Hydrate of iron.* (Brauneisenstein, Raseneisenstein, Lepidochrokite, Stilpnosiderite, &c.)

In the matrass they give off water, and leave a red oxide. The stilpnosiderite is fusible in a thin lamina, with a strong heat.

After solution in *salt of phosphorus*, it gives with tin, in a good reducing flame, some traces of copper.

16. *Manganese.*

1. *Sulphuret of manganese*, from Nagyag.[1]

Alone, in the matrass, no change.

In the open tube roasts very slowly without giving off any sublimate. The roasted surface assumes a clear green grey colour, and the interior remains a long time unchanged.

On charcoal, in a very good reducing flame, after roasting to a certain degree, the thin portions of the assay may be fused into a brownish scoria. When perfectly roasted, it behaves with the fluxes like pure oxide of manganese.

With borax fuses with great difficulty into a glass, which, when cold, has a slight yellowish tint,

---

[1] One atom of manganese, 28 + 1 of sulphur, 16 = 44. C.

as long as any a portion of the assay remains still untouched. This colour seems to be of the same kind as that which sulphur gives to the glasses of silica and soda. When the assay is dissolved, and the reduction ceases, the colour indicating oxide of manganese appears.

In salt of phosphorus dissolves with brisk effervescence, and abundant disengagement of gas, which continues some moments after the blast has ceased. If we then bring the assay globule near the flame of the lamp, we may hear little detonations occasioned by the inflammation of the combustible gas which is liberated. If the globule be large, it retains its heat so much the better, and the phenomenon is prolonged; at last a large bubble of gas bursts out, which burns with a pale green light. The disengaged gas is known, by the peculiar odour of the boiling matter, to be sulphuret of phosphorus, formed by the combination of the sulphur with the phosphorus of the phosphoric acid, which is disengaged from the latter by the oxidation of the manganese at the expense of the acid. In this operation the glass presents a peculiar play of colour. Whilst liquid it is transparent and colourless, but acquires on cooling the same yellow colour as glass of borax. Whilst the greater part of the mineral is still undecomposed, the mass by cooling becomes clear yellow, but afterwards its colour gets brown, and at the moment of congelation a fixed substance separates, which makes the globule black. With the microscope we see in the interior of the glass, whose proper colour is dark yellow, small black particles in suspension, and towards the end of the operation the light it transmits is bluish.[1] When the whole of the sulphuret of manganese is dissolved, and the effervescence has totally ceased, the glass becomes perfectly clear and colourless, and acquires in the oxidating flame a fine amethyst colour.

With soda, the solution is imperfect; a hepatic mass sinks into the charcoal, and a half fused grey scoria remains on the surface.

2. *Peroxide of manganese.*[2]

Alone, in the matrass, no perceptible change, if pure, but commonly the best crystallized peroxide of manganese contains more or less hydrate of manganese, whose water is expelled by heat, and its quantity serves as the basis to calculate the commercial value of the oxide; the more water the heated mass gives off, the less peroxide of manganese it contains, and the less it is worth.

On charcoal it becomes brown red in a good reducing flame.

In borax and salt of phosphorus dissolves with

---

[1] The cause of this last tint is certainly not derived from the sulphur. Can sulphuret of manganese be soluble in the salt of phosphorus, or may there be an oxide of manganese, of an inferior degree to that of the protoxide, and may this oxide, like the inferior oxides of copper and bismuth, be soluble in the fused acid, but liable to abandon it at the moment it becomes solid? B.

[2] One atom of manganese, $28 + 2$ atoms of oxygen, $16 = 44$. C.

brisk effervescence from the disengaged oxygen; in other respects it behaves like pure oxide of manganese, (p. 101). A peroxide of manganese is found in nature, containing a pretty large portion of iron, whose presence is ascertained by exposing the glass, particularly that with borax, to a good reducing flame. The characteristic colour of iron alone remains after the operation. It may also be detected by reduction with soda; the metal being reduced and the mass absorbed, it is collected from the charcoal by the usual process.

3. *Ferriferous phosphate of manganese,* from Limoges (Phosphormangan).[1]

Alone, in the matrass, gives off a little water, which acts on litmus paper as an acid, and turns brazil wood paper yellow; the transparency of the glass is not affected. But if we treat the assay *in the open tube,* directing the flame of the lamp into it, its sides here and there become opaque from the deposition of silica; the mineral therefore contains a little fluoric acid; yet the water which condenses in the tube does not act on brazil wood paper.

Alone, on charcoal, fuses very easily with brisk intumescence, into a black globule, with metallic lustre, and very magnetic.

[1] Vauquelin's analysis gave, oxide of manganese 42, oxide of iron 31, phosphoric acid 27. *Jameson.* Berzelius considers it to be composed of 1 atom of subphosphate of manganese, +1 atom of subphosphate of iron; both metals in the state of protoxide. C.

With borax fuses easily. The oxidating flame develops the characteristic colour of oxide of manganese, and the reducing that of iron.

With salt of phosphorus fuses very easily. Little else than the colour indicative of iron is perceptible; we can however produce a slight tinge of manganese by long oxidation assisted by a very low ignition.

With boracic acid fuses. By introducing a small bit of metallic iron into the mass, we obtain phosphuret of iron. (See p. 129.)

With soda, on charcoal, does not fuse, but in the reducing experiment gives a large quantity of phosphuret of iron. On the platina foil the usual action of oxide of manganese ensues.

4. *Carbonate of manganese,* from Freyberg.[1]

Alone, in the matrass, gives off a little water, and decrepitates violently. In a higher heat, carbonic acid is disengaged, and the assay becomes greenish grey, which distinguishes the protoxide of manganese. In a matrass with a large body, or on *charcoal,* it becomes black by the superoxigenation of the protoxide. On charcoal, in the reducing flame, it becomes brown black, and the effect of caustic lime is perceived by brazil wood paper. It afterwards behaves with the fluxes like slightly ferruginous oxide of manganese.

---

[1] A mixture of the carbonates of manganese, iron, and lime. C.

5. *Wolfram.*[1]

Alone, in the matrass, decrepitates a little occasionally, and gives off a small quantity of water.

On charcoal, in a good heat, may be fused into a globule, whose surface presents a collection of tolerably large lamellar, iron grey crystals, having a metallic lustre. If the assay be only partially fused, the crystallization on the surface is confused and indistinct.

With borax fuses readily, and gives the characteristic colours of iron, without enabling us to discover from their play the presence of tungstic acid.

With salt of phosphorus fuses easily. In the oxidating flame the glass assumes only the tint of iron; but in the reducing flame it becomes dark red. A very small proportion of wolfram renders it opaque, an exceedingly minute quantity producing an evident effect. If we add tin, and blow an instant, the colour becomes green; but this is not well distinguished if it be too intense, for in that case the colour is so deep that the globule usually appears opaque. Wolfram is completely reduced in a good, long continued reducing flame, and the green colour disappears, and is replaced by a permanent feeble reddish tint.

---

[1] Vauquelin's analysis gave, tungstic acid 67, oxide of manganese 6·25, oxide of iron 18·10, silica 1·5. *Jameson.* Berzelius considers it a compound of 1 atom of tungstate of manganese, $(36+120)$ $156+3$ atoms of tungstate of iron, $(36+120 \times 3)$ $468 = 624$. C.

With soda, on platina foil, decomposes and falls to powder. The soda is coloured green by manganese on the edges. On charcoal it is easily reduced into an alloy of tungsten and iron, which may be separated from the charcoal by washing, &c.

Remark.—Wolfram sometimes decrepitates very strongly, and splits into thin leaves. In that case it is usually covered with an earthy, somewhat hard, yellow coating, which might easily be mistaken for tungstic acid, but is in reality arseniate of iron, and the wolfram, when acted on by the reducing flame, disengages a strong odour of arsenic.

6. *Columbite.*

(*a*). *Columbite*, from Kimito in Finland.[1]

Alone, no change.

With borax dissolves slowly, but perfectly. The glass presents only the tint of iron; at a certain point of saturation it takes by *flaming* a grey white colour, and when still further saturated, it spontaneously becomes opaque on cooling. As long as it remains transparent its colour is pale bottle green.

---

[1] According to Dr. Wollaston's analysis it consists of oxide of columbium 85, oxide of iron 10, oxide of manganese, 4; and that from America, of oxide of columbium 80, oxide of iron 15, oxide of manganese 5. *Phil. Trans.* 1809. Berzelius considers it as composed of 1 atom of columbate of manganese (36+152), 188+1 atom of columbate of iron 36+152) 188 = 376. C.

With salt of phosphorus it dissolves slowly, and has merely the colour indicative of iron. *In the reducing flame* the glass does not become red on cooling, whence it is clear it contains no tungstic acid.

With soda, on platina foil, the action of manganese is evident. *On charcoal,* if we add a little borax, which dissolves the columbiferous compound, and prevents the reduction of the oxide of iron, we obtain, by the usual reducing process, a small quantity of tin.

(*b*). *Columbite,* from Broddbo.[1]

Alone, no change.

With borax behaves likes the preceding.

With salt of phosphorus dissolves slowly; assumes in the oxidating flame the colour of iron, and, in the reducing, a red colour, which increases on cooling, and denotes tungsten. The addition of tin does not alter this colour, nor produce in the present instance the green colour, which pure tungsten gives in similar circumstances.

With soda, behaves like the preceding, but

---

[1] By one analysis Berzelius found it to consist of oxide of columbium 66·66, tungstic acid 5·78, oxide of tin 8·02, oxide of iron 10·64, oxide of manganese 10·20. These ingredients varied a little in two other analyses, which afforded nearly 2 per cent. of lime. *Phillips.* By the formula, Berzelius seems to consider it as composed of 1 compound atom of columbate of lime, and columbate of iron, +3 compound atoms of columbate of manganese and columbate of iron, +1 atom of wolfram, and an atom of tin. C.

with soda and borax, in the reducing flame, gives a larger quantity of tin.

(c). *Columbite,* from Finbo.

Behaves like that from Finland, but in the reducing experiment gives a considerable quantity of tin. The proportion of tin appears to be variable, and some specimens are obviously nothing else but columbiferous ores of tin. The presence of columbium in these ores is always detected by the glass which they give with borax, being more or less tinged by iron, and assuming the appearance of enamel by *flaming.*

(d). *Columbite,* from Bodenmais.[1]

Alone, no change.

With borax fuses into a black or very dark, almost opaque bottle green glass. It does not become opaque by *flaming,* till it has acquired so deep an iron tint as to be incapable of transmitting light. This serves to distinguish columbite with excess of base, from neutral columbite.

With salt of phosphorus fuses slowly into a glass deeply coloured by iron, in which we can discover no trace of tungsten.

With soda, on platina foil, the action of manganese is evident; with *soda and borax,* we obtain by the reducing flame some traces of tin.

---

[1] By the analysis of Vogel it consists of columbium 75, protoxide of iron 17, protoxide of manganese 5, oxide of tin 1. *Phillips.* By Berzelius's formula it is composed of 1 atom of sub-columbate of manganese +3 atoms of sub-columbate of iron. C.

(c). *Columbite*, from Connecticut.[1] (Haddaw.)

With borax behaves like the columbite from Bodenmais, and with *salt of phosphorus* like that from Broddbo. It appears, therefore, to be a tungsteniferous columbite, with excess of base.

7. *Black mangankiesel*, from Klapperud near Dal.[2]

Alone, in the matrass, gives a large quantity of water free from acidity, and afterwards an empyreumatic gas loaded with vapour, whilst the black colour of the assay changes to clear grey. Heated to redness, it intumesces, and its colour becomes still clearer.

On charcoal intumesces, and fuses into a glass, which is bottle green in the reducing flame, but blackens, and assumes a metallic lustre in the oxidating.

With borax fuses readily into a glass, which in the exterior flame is strongly tinged by manganese, and in the reducing flame receives a slight tint from protoxide of iron.

With salt of phosphorus, fuses into a colourless glass, which becomes amethyst red in the exterior flame, and leaves a skeleton of silica.

With a small quantity of soda fuses into a black glass; a larger proportion of soda gives a

---

[1] I have found this columbite in the large grained granite containing cymophane. Its gangue is albite, precisely like that from Finbo. B.

[2] By the formula this is a hydrate of subsesqui silicate of manganese. C.

black scoria, and the flux is absorbed by the charcoal.

8. *Red mangankiesel* (Rubinspat), from Longbanshytta, in Wärmelande.[1]

Alone, in the matrass, no change.

On charcoal no charge till it begins to fuse, when it gives, in the reducing flame, a semi-transparent glass of the same colour as the mineral; but in the oxidating flame, it forms a black globule, with metallic brilliancy, whose colour may be made to disappear by the reducing flame.

With borax, in the reducing flame, readily fuses into a colourless glass: in the oxidating into one of an amethyst colour.

Salt of phosphorus attacks it with difficulty. The result of its action is a skeleton of silica, and a colourless glass, which in the exterior flame assumes an amethyst colour.

With a little soda fuses into a black glass; a larger quantity of the flux forms a black scoria, of difficult fusion, and a still larger quantity causes it to pass into the charcoal.

9. *Pyrosmalite,* from the iron mine at Nordmark.[2]

---

[1] Bisilicate of protoxide of manganese, by the formula. C.

[2] Analysis by Hisinger; protoxide of iron 21·810, protoxide of manganese 21·140, submuriate of iron 14·095, silica 35·850, lime 1·210, water and loss 5·895. *Phillips.* By the formula, it consists of 1 atom of bisilicate of protoxide of manganese +1 atom of bisilicate of protoxide of iron, in mechanical mixture, as Berzelius supposes, with 14 per cent. of submuriate of iron and water. C.

Alone, in the matrass, at first gives it off water; by increasing the heat, we obtain a yellow substance, (muriate of iron,) which dissolves into little drops of the same colour in the last portions of the water; they redden litmus paper, and have an astringent flavour.

On charcoal, with a gentle heat, pyrosmalite exhales a weak acid odour, fuses readily into a globule, with a brilliant smooth surface and iron grey colour.

With borax fuses readily, and exhibits the characteristic colours of iron.

With salt of phosphorus the solution is more difficult, and the interior of the glass continues for a long time obscure; at last the assay fuses and acquires the iron tint, leaving a skeleton of silica.

With nitre it exhibits a strong tint of manganese.

With soda, on platina foil, it presents the effects of manganese. *On charcoal* it gives a black glass, which, with a large quantity of soda, sinks into the charcoal.

If we treat a morsel of pyrosmalite with solution of oxide of copper in salt of phosphorus, the clear blue aureola, characteristic of muriatic acid, forms round the globule; but the phenomenon is momentary.

10. *Silicate of manganese,* from Piemont.[1]

[1] See "Nouveau Système de Minéralogie," by Berzelius, Paris, 1819, p. 277. B. In the note referred to is the result of the analysis of this mineral by the author; it gave him

Alone, in the matrass, no change in its appearance.

On charcoal, in a very strong heat, fuses on the edges and preserves its grey black colour.

With borax fuses readily. In the exterior flame it exhibits an amethyst colour, and in the interior a slight tint of protoxide of iron.

In glass of phosphorus dissolves readily, effervesces, and develops a lively amethyst colour. In the reducing flame, the glass becomes colourless, presents internally a residuum of silica, uniformly diffused through the globule, and is very opaline on cooling.

With soda no solution.

11. *Hydrate of manganese,* crystallized from Undenäs. *Earthy hydrate* (Wad), from various places.

Alone, in the matrass, gives off abundance of water.

On charcoal, and *with the fluxes,* behaves like oxide of manganese.

12. *Kupfermangan,* from Schlackenwald.

Alone, in the matrass, first gives off a large quantity of water, then decrepitates and splits to pieces. The water is not acid.

---

silica 15·17, chesnut brown oxide of manganese (composed, according to Hisinger, of 1 atom of protoxide + 2 atoms of peroxide), 75·80, alumina 2·80, oxide of iron 4·14. He thinks its probable composition is that of a subsilicate, containing 3 atoms of peroxide of manganese +1 atom of silica. C.

On charcoal, in the reducing flame, becomes brown, but does not fuse.

With borax dissolves readily, and assumes the manganese tint. In the reducing flame it gives a transparent colourless glass, which becomes red and opaque on cooling.

With salt of phosphorus fuses readily, and exhibits the same shades as with borax. If the colourless glass, obtained by the reducing flame, be exposed for a moment to the oxidating, it assumes the fine green tint of copper, and retains its transparency on cooling. If the oxidation be carried further, it acquires a bluish amethystine colour.

With soda, no solution; but, if we add a little borax, we obtain very evident fused globules of copper.

17. *Cerium.*

1. *Fluate of cerium* from Finbo, Broddbo, and Bastnäs.

(*a*.) *Neutral fluate*, from the two first places.

Alone, in the matrass, gives off a little water. At the temperature at which glass begins to melt, a slight corrosive action is perceptible on the substance of the matrass, at a little distance from the assay. The water colours brazil wood paper yellow, and the yellowish colour of the assay changes to white.

In the open tube, the flame being thrown into the tube, its sides are acted on and rendered opaque by the silica which is deposited on them. The water condensed in the tube yellows brazil wood paper. The assay becomes yellow brown.

On charcoal does not fuse; the colour of the mineral merely turns a little brown.

With borax, and salt of phosphorus, the effects are the same as those of oxide of cerium. (See page 100.)

With soda, fluate of cerium splits and intumesces, but does not fuse; the soda is absorbed by the charcoal, and leaves a grey mass on the surface.

(*b.*) *Fluate with excess of base,* from Finbo.

Alone, in the matrass, gives off water, and its colour becomes brown.

On charcoal, changes colour on being heated, and appears black at a temperature bordering on incipient redness; but, on cooling, it becomes dark brown, then fine red, and lastly orange. This change of colour at once distinguishes it from the neutral fluate, which does not offer the same phenomenon. It is infusible. In other respects it behaves with the fluxes like the preceding, except that its cohesion is not destroyed by soda, and that it remains entire, at least if the blast be not too strong, nor too long continued.

(*c.*) *Fluate of cerium,* from Bastnäs (not as yet analysed).

Alone, in the matrass, gives off a little moisture, without any change in its appearance.

On charcoal does not fuse; turns opaque at a low heat; becomes by the fire rather darker, and presents the same changes of colour as the preceding.

In the open tube, under the *immediate* action of the flame, it gives very decided indications of fluoric acid.

With borax, and salt of phosphorus, like the preceding.

With soda, neither loses its cohesion, intumesces, nor dissolves.

Remark.—This mineral appears to be a subsalt, with perhaps a slighter excess of base than in the preceding.

2. *Cerite*, from Bastnäs, near Riddarhytta.[1]

Alone, in the matrass, gives off water, and becomes perfectly opaque.

On charcoal splits here and there, but does not fuse.

With borax the solution is effected slowly. In the oxidating flame we obtain a deep orange coloured glass, whose tint gets clearer on cooling, and at last becomes clear yellow; by *flaming*, the glass assumes the appearance of white enamel, and a slight iron tint in the reducing flame.

Salt of phosphorus combines with the oxide of cerium, and developes its distinguishing colours. The glass is colourless when cold, and the silica remains suspended in it, in the form of a white opaque skeleton.

With soda does not dissolve, but fuses imperfectly into an orange coloured scoriaceous mass.

---

[1] Vauquelin's analysis gave oxide of cerium 67, silica 17, oxide of iron 2, lime 2, water and carbonic acid 12. *Jameson*. According to the formula, Berzelius considers it as a silicate of protoxide of cerium, composed of 3 atoms of base + 2 of silica, with 6 atoms of water. *C*.

DIVISION II.

METALS NOT REDUCIBLE BY CHARCOAL, AND WHOSE OXIDES FORM EARTHS AND ALKALIES.

1. *Zirconium.*

Zircon and *hyacinth*, from Ceylon, Finbo, Fredriksvärn, and Expailly.

Alone, colourless, transparent zircon suffers no change. The red transparent zircon (hyacinth) loses its colour, and becomes either perfectly limpid, or very slightly yellow. The brown opaque zircon from Fredriksvärn loses its colour and becomes white, and like glass that is full of cracks (fendillé). The blackish zircon from Finbo gives off a little moisture, becomes milk white, and has an effloresced appearance. None of these varieties are fusible, either in powder or thin laminæ.[1]

With borax zircon fuses readily into a diaphanous glass, which, at a certain point of saturation, is capable of becoming opaque by *flaming*, and when still further saturated, becomes obscure spontaneously on cooling.

Salt of phosphorus does not attack zircon. A fragment submitted to its action retains all the sharpness of its edges; and even when the assay has been previously pulverised, we cannot discover any effect of the flux. The glass remains perfectly colourless (or milky white, if the assay be pulverised), both in the oxidating and reducing flame, so that

---

[1] "Exposed to the powerful heat of the *gas blowpipe* zircon becomes first opaque, and of a white colour; afterwards, its superficies undergoes a partial fusion, and exhibits a *white opaque enamel*, resembling *porcelain*." (*Clarke.*) C.

no trace of titanium which Chevreul found in the zircon of Ceylon, and John in that from Norway, is perceptible.[1]

Soda does not dissolve it; it attacks it slightly on the edges, and then penetrates into the charcoal. On platina foil most of the zircons give traces of manganese.

2. *Aluminium.*

1. *Telesia.* Ruby. Sapphire. Corundum.

Alone no change, whether in fragment or powder.[2]

With borax fuses slowly, but perfectly, into a transparent colourless glass, which cannot be made opaque by flaming.

---

[1] In examining those rolled grains, similar to hyacinth, which we meet with in collections under the name of *Adelstensgrus* (gem gravel), the greatest part of which is spinel, I found some that had every appearance of hyacinths, which dissolved in part in salt of phosphorus, and then exhibited a slight tint of titanium. It is possible that a mixture of this kind of grains, with the zircon of Ceylon, may have led to the idea, that that mineral contains titanium. With respect to John's result, it is clear from his own description, that what he took for titanium was not that substance.

To obtain zircon from the gem sands of Ceylon, it may perhaps be necessary first to heat the zircons and hyacinths to redness, and then to select the colourless grains, for those which retain their colour are spinels, essonites, or pyropes. B.

[2] By the gas blowpipe, Dr. Clarke fused two rubies into one bead; they lost their red colour in the operation. Pure crystallized blue sapphire was also fused, "and exhibited, during fusion, the singular appearance of greenish glass balloons, swelling out in grotesque forms, which remained fixed when the mineral became cool." Common corundum fused, but with difficulty, into a greenish translucid glass; the fusion was attended with a slightly coloured greenish flame. See Gas Blowpipe, p. 54, et seq. C.

With salt of phosphorus fuses slowly (it must be in powder) and gradually into a transparent glass. If more of the assay have been added than the flux can dissolve, the undissolved portion does not, as with the silicates, become transparent; nor does the glass become opaline, either by cooling or by the action of the exterior flame.

With soda, no action, nor appearance of fusion.

With solution of cobalt we obtain a dark blue; the more perfectly the assay has been pulverised, the stronger the heat, and the better it is kept up, the finer is the colour, for the oxide of cobalt acts with considerable difficulty on alumina in this state.

2. *Aluminite,* alumina from Halle.[1]

Alone, in the matrass, gives off a great deal of water, and, at incipient redness, sulphurous acid, distinguishable by its odour, and by its action on moistened brazil wood paper.

On charcoal, and with *the fluxes,* it behaves like alumina. The aluminite, from Newhaven, deposits some flakes of silica in its solution by phosphoric acid.

With solution of cobalt we obtain a bright blue colour, pleasant to the eye.

3. *Wavellite,* from Barnstaple, Annaberg in the Palatinate, and from Bohemia.[2]

---

[1] Subsulphate of alumina, with water. C.

[2] Berzelius' very masterly analysis of this mineral gave him, alumina 35·35, phosphoric acid 33·40, fluoric acid 2·06, lime 0·50, oxides of iron and manganese 1·25, water 26·80. He

Alone, in the matrass, gives off water, the last drops of which are acid, have a gelatinous consistence, in consequence of the silica they contain, and colour brazil wood paper yellow. When evaporated they leave a deposit of silica on the glass, which disturbs its transparency. Little circles of silica form above the assay during ignition.

On charcoal it intumesces, loses its crystalline form, and becomes snow white.

With borax, salt of phosphorus, soda, and *solution of cobalt,* it behaves like aluminite.

Treated with boracic acid and iron, in the manner described p. 129, it gives a fused regulus of phosphuret of iron.

4. *Lazulite.* Blue feldspar, from Krieglach.[1] Phosphate of alumina in an unknown degree of saturation, mixed with phosphate of magnesia and phosphate of protoxide of iron.

Alone, in the matrass, gives off water, and loses its colour.

On charcoal intumesces, and, at the point where the heat is greatest, assumes a vitreous, blebby aspect; does not fuse.

---

considers it to be a subphosphate of alumina with water, probably mechanically mixed with a small quantity of neutral fluate of alumina. Nouveau Système, p. 278. C.

[1] The blue mineral from Vorau differs a little from this in the phenomena it exhibits. It intumesces much more, and falls to pieces; *gives no appearance of fusion,* nor any blue colour with solution of cobalt, till after having *entered into fusion* (literal); the blue it gives then is sensibly reddish. B.

With borax fuses into a transparent colourless glass.

With salt of phosphorus, first becomes transparent on the edges, and gradually fuses entirely into a colourless diaphanous glass.

With soda intumesces, but neither dissolves nor fuses.

With boracic acid and iron gives phosphuret of iron.

With solution of cobalt produces a fine blue colour.

5. *Calaïte*, Persian turquoise. A mixture of phosphate of alumina with phosphate of lime and silica, coloured green, or bluish green, by carbonate and hydrate of copper.[1]

Alone, in the matrass, gives off a little water, and decrepitates with much violence, even though very slowly heated. The water has no effect on litmus paper. After splitting, the assay is black.

On charcoal, or in the forceps, becomes brown in the interior flame, and tinges its point green. Does not fuse, but assumes a vitreous appearance

---

[1] In these experiments I used a mamellary blue turquoise, and a green turquoise; the latter was given me by the Hon. Mr. Strangways, as a specimen of the true calaïte, and exactly answers its description. It is a thin lamina covered on both sides with a grey argillaceous substance. Observing that it gave the characteristic indications of phosphoric acid, before the blowpipe, I analysed it in the humid way, and found in it phosphate of alumina, phosphate of lime, silica, oxide of iron, and oxide of copper. Consequently, John was mistaken in calling calaïte a hydrate of silica. B.

on the surface, at the point where the heat acted most powerfully.

With borax dissolves readily into a limpid glass, which exhibits the tint indicative of iron whilst hot, and assumes, when cold, a slight copper green tinge, if it has been exposed to the exterior flame; or, becomes red and opaque, if to the interior, especially if tin has been employed.

With salt of phosphorus fuses easily and perfectly into a transparent glass, which exhibits the same alternations of colour as the glass with borax.

With a small quantity of soda, at first intumesces, and then slowly fuses into a semi-transparent glass, coloured by iron. With a larger quantity of soda it becomes infusible; if we add still more soda, we obtain a considerable portion of copper in the reducing flame.

With boracic acid and iron we obtain a regulus of phosphuret of iron.

6. *Topaz.*[1]

Alone, in the matrass, no change, nor trace of fluoric acid.

On charcoal does not fuse.[2] The yellow topaz,

---

[1] Vauquelin's analysis of Brazilian topaz gave him, silica 29, alumina 50, fluoric acid 19, loss 2. *Jameson.* Berzelius considers it to be composed of 1 atom of sub-fluate of alumina + 3 atoms of silicate of alumina. Klaproth's analysis gave much more silica than Vauquelin's. C.

[2] By the gas blowpipe, Dr. Clarke fused topaz into a white enamel, covered with minute limpid glass bubbles. *Gas Blowpipe,* p. 59. C.

by a low degree of ignition, becomes pale rose colour, in consequence of the conversion of the hydrate of iron into peroxide. This topaz, as well as the colourless limpid variety, retains its transparency. In a very strong heat, the longitudinal faces of the crystal are covered with numerous small white bubbles, which give them a frosted appearance, but are not easily discerned without the microscope; they are not perceptible on the lamellar surface of the transverse section. These little bubbles are not so distinct in the opaque topaz from Finbo and Brodbo, but they may be enlarged by a good heat to a certain extent, beyond which they usually burst. A very high temperature is requisite to produce this effect, and the experiment succeeds only with small fragments.

With borax fuses slowly into a transparent glass. The limpid topaz becomes white and opaque before it dissolves.

With salt of phosphorus dissolves slowly, and leaves a silica skeleton; the globule is transparent, and becomes opaline on cooling.

With a small quantity of soda it is converted by laborious solution into a colourless, semi-transparent, blebby scoria. With a larger quantity of soda, it swells up and becomes infusible.

With solution of cobalt it gives a blue colour, impure and unpleasant to the eye.

7. *Pycnite,* from Altenberg.[1]

[1] Composed of 1 atom of fluate of alumina + 3 atoms of silicate of alumina. Dr. Clarke fused it into a snow white enamel.

Behaves like topaz, except that when heated by itself, the bubbles are produced more easily and in larger quantity.

8. *Disthene*, from St. Gothard, and Norway (Cyanite, Sappare), and Rhœtizite, from the Tyrol.[1]

Alone, no change at a red heat; in a very strong heat, whitens without fusing; even its powder is infusible. Rhœtizite becomes red at a low heat, but in a more intense one it turns white.

With borax fuses slowly, but perfectly, into a transparent colourless glass.

With salt of phosphorus dissolves in part, and leaves a blebby, semi-transparent silica skeleton. The globule does not become very sensibly opaline by cooling.

With a small quantity of soda fuses imperfectly into a blebby, semi-transparent, rounded mass. If fused in the exterior flame, the assay assumes a pale rose colour, and becomes transparent in the coloured part. Exposed to an intense heat in the interior flame, this colour disappears, and cannot be reproduced by the exterior flame. It is developed better on the platina wire than on charcoal, by means of a pretty large quantity of soda, for the charcoal absorbs the soda before it can act on the assay. The red colour is much more apparent in the cyanite from St. Gothard than in that from Norway; the glass formed with the rhœtizite merely becomes yellowish. A larger proportion of

[1] A subsilicate of alumina. Before the flame of the gas blowpipe, disthene fused readily into a snow white frothy enamel. *Clarke.* C.

soda causes the assay to intumesce, and renders it infusible.

With solution of cobalt, in a strong heat, it assumes a fine deep blue colour.

9. *Nepheline* (primitive), from Vesuvius.[1]

Alone, on charcoal, its edges become rounded, without any sensible intumescence; cannot be fused into a globule, but gives a blebby, colourless glass.

With borax fuses slowly, and without effervescence, into a transparent colourless glass.

With salt of phosphorus dissolves without effervescence, and leaves a silica skeleton. The glass globule becomes opaline on cooling.

With soda, at first intumesces, then fuses into a blebby colourless glass.

With solution of cobalt the unfused portion of the pulverulent mass has a greenish grey colour; the fused edges are grey blue.

Remark.—It is evident from these experiments that cyanite and nepheline differ in their composition, although the analysis of the first by Klaproth, and that of the second by Vauquelin, give nearly the same result.

10. *Pinite*, from St. Pardou, in Auvergne, and from Greenland.[2]

---

[1] According to the mineralogical formula annexed, it consists of 1 atom of silicate of soda + 3 atoms of silicate of alumina. C.

[2] The last was given me by M. Haüy. It is crystallized in the form of a hexahedral prism, and purer than the former. Pinite may, in general, be said to behave like fusible crystallized alumina. B.

Alone, in the matrass, gives off a little water, without any change in its appearance.

On charcoal whitens, and fuses on the edges into a white blebby glass. The pinite from Auvergne becomes covered with coloured spots. The most ferruginous variety sometimes fuses readily into a black glass.

With borax the solution is extremely difficult, even if the assay be pulverised. The result is a transparent glass faintly tinged with iron.

Salt of phosphorus has no visible action on a fragment of the Auvergne pinite, but the bead exhibits a tint of iron as long as it is hot. When pulverised, it is decomposed by a prolonged heat, and leaves a residuum of silica. The glass becomes opaline on cooling. The pinite from Greenland is more easily decomposed by salt of phosphorus; in other respects its glass exhibits the same properties.

With soda dissolves slowly into an opaque glass, slightly coloured by oxide of iron, and of difficult fusion.

11. *Fahlunite,* from the mine of Eric-Matts, at Fahlun.

Alone, in the matrass, gives off water free from acidity.

On charcoal whitens and fuses on the edges into a white blebby glass.

With borax fuses slowly into a glass slightly coloured by iron.

With salt of phosphorus decomposes, and leaves a skeleton of silica; the glass is transparent, and tinged by iron whilst hot, but becomes opaline and colourless on cooling.

With soda does not dissolve, but assumes the appearance of scoria, and becomes tinged of a yellowish colour.

12. *Allophane,* from Saxony, and Thuringia.

Alone, in the matrass, gives off slightly acid water, and becomes sprinkled with black spots. The substance is perceived to be heterogeneous, and traversed by white testaceous laminæ, which are not coloured.

On charcoal, or *in the forceps,* it does not fuse, but intumesces, and readily falls to powder, and indicates the presence of copper, by the green colour it imparts to the flame.

With borax fuses very slowly into a colourless glass, which in the interior flame assumes a pale red tint, and with the help of a little tin, a dark red colour, from protoxide of copper.

Salt of phosphorus readily decomposes allophane, leaves a silica skeleton, and exhibits slight tints of the colours indicative of copper. Tin colours the glass red.

Soda does not dissolve it; the assay becomes green in the oxidating flame, and red in the reducing. By the addition of borax, we may extract globules of metallic copper.

13. *Carpholite,* from Schlackenwald, in Bohemia.

Alone, in the matrass, gives off water, which, by igniting the assay, is acid, attacks the glass, and yellows brasil wood paper. The sides of the matrass are covered here and there with silica, deposited by the fluoric acid.

On charcoal first intumesces, then whitens, and fuses slowly into a brown opaque glass, which becomes much darker in the exterior flame than in the interior.

With borax fuses into a transparent glass, which assumes the colour of manganese in the exterior flame, and becomes greenish in the interior.

With salt of phosphorus intumesces, and leaves a glassy skeleton of silica, which, contrary to its usual habit, dissolves with the greatest ease into a transparent glass, that becomes very opaline on cooling. In the oxidating flame it assumes a sensible amethystine colour.

With soda does not dissolve *on charcoal.* The assay intumesces, and assumes a fine green colour. *On platina foil*, on the contrary, it dissolves in a sufficient quantity of soda, and forms a dark green mass easily fusible.

With solution of cobalt assumes a rather impure dark blue colour.

With boracic acid and iron exhibits no trace of phosphoric acid.

14. *Staurotide*, from Saint Gothard.[1]

[1] Klaproth made two analyses of this mineral, in which the proportions of the alumina and silica are sufficiently discordant. In one the alumina is stated as 52·25 per cent., in the other as 41; and the silica in the former as 27, in the latter 37·5. The oxide of iron is very nearly the same in both, 18·25. *Jameson.* From one of these analyses, which he

Alone, in fragment, infusible, and suffers no change, except that its colour becomes darker, and almost black. In powder it fuses on the edges into a black scoria.

With borax fuses slowly into a transparent dark green glass, coloured by protoxide of iron.

With salt of phosphorus, unless the assay be in powder, the fusion proceeds extremely slowly. Little or no silica is left in the solid form; the glass whilst still warm is transparent, and yellowish green, but becomes opaline and loses its colour on cooling.

Soda does not dissolve it, but combines with it with effervescence, and forms a yellow scoria.

With solution of cobalt it does not become perfectly blue, but the fused parts assume a dark colour, inclining to dirty blue.

15. *Almandine, argillaceous garnet.*[1]

Alone, becomes brown by heat, but recovers its natural colour on cooling; fuses without the slightest intumescence into a black globule, with an unpolished, metallic surface, and appearing as if covered with a pellicle of reduced iron. Whilst cooling, a cavity forms somewhere on the globule, from the contraction of its parts. Its fracture is vitreous.

With borax fuses very slowly into a dark glass

---

alludes to in the Nouveau Sytème, Berzelius calculates the composition of Staurotide, which, from the formula, he considers as composed of 1 atom of sub-silicate of iron + 6 atoms of sub-silicate of alumina. C.

[1] By the formula, composed of 1 atom of silicate of iron + 1 atom of silicate of alumina. C.

tinged by iron. The solid nucleus inclosed in the glass appears dark.

Salt of phosphorus decomposes it, occasions intumescence, and the formation of a skeleton which at first is white, spotted with black, but becomes colourless by a prolonged blast.

As long as the skeleton retains its cohesion, the glass is transparent after cooling; but if we continue to blow till it fuses, the globule becomes opaline on cooling.

By soda, almandine is decomposed, and intumesces, then fuses into a black globule with a metallic lustre. An increased quantity of soda does not diminish its fusibility. On platina foil, we discover traces of manganese.

16. *Garnet,* from Finbo.[1]

Alone, on charcoal, behaves like the preceding, except that the surface of the fused globule is metallic only in some places.

With borax fuses like almandine, but in the oxidating flame, the saturated glass assumes the colour of amethyst.

With salt of phosphorus behaves like almandine.

Soda decomposes it; with a small quantity of that flux it fuses slowly into a black globule; a larger dose lessens its fusibility, and at last entirely

---

[1] Its composition, by the formula, is supposed to be 1 atom of bisilicate of protoxide of iron + 1 atom of silicate of protoxide of manganese + 2 atoms of silicate of alumina. C.

destroys it. On platina foil it exhibits the effects of manganese very distinctly.

17. *Garnet*, from Broddbo.[1]

Alone, fuses with slight bubbling into a black globule, whose whole surface shines with vitreous brilliancy.[2]

With borax behaves like the preceding.

With salt of phosphorus also like the preceding; but the globule does not so readily become opaline on cooling.

With soda, like the preceding.

18. Diaspore.[3]

---

[1] The formula states its composition as 1 atom of bisilicate of protoxide of iron + 2 atoms of silicate of protoxide of manganese + 2 atoms of silicate of alumina. C.

[2] The aluminiferous garnets (*lergranater*), appear to be double silicates of alumina, and one of the four bases, protoxide of iron, protoxide of manganese, magnesia, and lime; sometimes of two of them, or even of all together. If the protoxide of iron predominate, the garnet is covered, after fusion, with a pellicle of reduced iron; but in proportion as it contains a larger quantity of the other bases, the iron coating is thinner, and disappears, and the surface of the globule becomes vitreous. In this way we can, by the blowpipe distinguish the calcareous garnet (aplome), from the garnet containing protoxide of iron (almandine). The garnets, Nos. 16 and 17, are mixtures of garnet with protoxide of iron, and garnet with protoxide of manganese. B.

[3] This mineral is as yet known only by an unique specimen which M. Le Lievre met with at a dealer's. M. Haüy, who has described it, had the goodness to give me a small piece of this substance, for experiment, taken from the specimen he received from M. Le Lievre. According to Vauquelin's analysis, it must be a hydrate of alumina; but the effects I obtained, show that it contains, besides, an alcaline element.

Alone, in the matrass, decrepitates violently, and splits into small white brilliant scales. During the decrepitation it gives off but little water, but when it is nearly red hot, it affords a pretty large quantity; hence it retains the water with considerable force, resembling in that respect most of the hydrates. If after we have heated the scaly fragments of the assay to slight redness, we place them on reddened litmus, or brazil wood paper, moistened with water, each scale forms a blue stain on the spot where it lay, and around it.

On charcoal the small scales are infusible.

With borax the scales fuse readily into a colourless glass, which does not become opaque by *flaming*.

With salt of phosphorus they fuse pretty readily into a colourless glass, without any siliceous residuum.

Soda has no action on them.

With boracic acid and iron they give no indications of phosphoric acid.

With solution of cobalt they assume a fine blue colour.

19. Clays.

(*a.*) *Fullers' earth*, from England.

Alone, in the matrass, gives off water, and becomes clear at first, but afterwards turns brown, and exhales an empyreumatic odour. The disengaged water acts feebly like solution of ammonia.

B. Vauquelin's analysis gave, alumina 80, water 17, iron 0·3. *Phillips.* There is a small specimen of this very rare substance in the British Museum. C.

On charcoal it crackles, and even splits with violence, unless very gradually heated. By a continued heat it whitens, and fuses into a white blebby glass.

Borax dissolves it slowly, so that the last portions require for the fusion a prolonged blast. The glass is transparent and colourless.

With salt of phosphorus it fuses, leaving a silica skeleton, into a transparent glass, which, after a brisk blast, becomes opaline on cooling.

With soda fuses into a glass globule of a bottle green colour.

With solution of cobalt it blackens.

(*b.*) *Cologne clay.*

Alone, in the matrass, it gives off water like the preceding, but the water is not alcaline.

On charcoal, or rather *in the forceps*, it must be heated gradually if we wish to prevent its splitting; it then fuses in the thin parts with a strong heat into a white glass.

With borax, salt of phosphorus, and soda, behaves like the preceding.

With solution of cobalt, it gives an impure blue.

(*c.*) *Stourbridge clay,* clay from Rouen, and from the coal mines of Höganäs; commonly called *apyrous clays.*[1]

In the matrass they behave like the preceding.

---

[1] Trisilicates of alumina. The formula is calculated from Mr. Sefström's analysis of the clays from Stourbridge, and Helsinborg, in the Annales du Bureau des Fers, (Jern-contoirets Annaler), for the year 1820. B.

On charcoal, with a gentle heat, they lose their deep brown colour, and become white. By a strong heat they are converted in the thin parts into a colourless glass having the appearance of a scoria, but do not fuse into a globule.

With borax and salt of phosphorus they behave like the preceding.

With soda they fuse into a transparent glass, slightly coloured by iron.

With solution of cobalt we obtain a pale blue.

3. *Yttrium.*

1. *Fluate of yttria and cerium*, from Finbo. A mechanical mixture of fluate of yttria, and fluate of cerium, with silica, either in the state of silicated fluoric acid, or as a mechanical aggregate.

It behaves like the neutral fluate of cerium (p. 209), except that we may add a large quantity of it to the glass of borax, before it acquires the property of becoming opaque by *flaming*. The most siliceous earthy fluates give with soda a coherent scoriaceous mass, on which a further addition of soda produces no change.

2. *Yttro-columbite*, from Ytterby, and Finbo. *Black and yellow*, a mechanical mixture of sub-columbate of yttria, with small quantities of the sub-columbates of lime and uranium, and sometimes with columbite and tungsten. *Dark* yttro-columbite, a tri-sub-columbate of yttra, mixed with the same substances.

Alone, in the matrass, gives off water, and the

black variety becomes yellow; some specimens become spotted, from containing black particles unalterable by the heat. In a red heat it whitens, and the substance of the matrass above the assay is acted on; the water, which is disengaged at the same time, at first turns brazil wood paper yellow, and then bleaches it.

With borax, fuses into an almost colourless glass, which, in a certain state of saturation, may be made opaque by flaming, and when still more saturated, becomes opaque spontaneously.

With salt of phosphorus at first is decomposed; the oxide of columbium retains the solid state, in the form of a white skeleton, but is fusible by a prolonged blast. The black yttro-columbite, from Ytterby, gives a glass which, after exposure to a good reducing flame, acquires on cooling a slight rose colour from the presence of a certain quantity of tungsten. The dark and yellow varieties from Ytterby assume on cooling a fine pale green tint, from a portion of uranium which they contain. In the yttro-columbite from Finbo, and Korarf, the effect of the uranium is concealed by a very intense colour derived from iron.

With soda decomposes without dissolving. *On platina foil* the effect of manganese is visible. By reduction with *soda and borax* we obtain from some varieties particles of tin, but that from Finbo contains so much iron, that the tin cannot be perceived.

3. *Gadolinite* A silicate of yttria.

(*a.*) *Gadolinite*, from Korarf. A mixture of gadolinite with small quantities of bisilicate of lime, and the silicates of protoxide of manganese, cerium and iron, and silicate of glucine.[1]

Alone, in the matrass, gives off a little water.

On charcoal whitens, and, in a good flame, fuses quietly into a dark or reddish pearl grey glass.

With borax fuses readily into a transparent glass, very slightly coloured by iron. If we saturate the vitreous globule, it becomes opaque, crystallizes on cooling, and assumes a grey colour inclining to red or green, according to the degree of oxidation of the iron; but it does not, like pure yttria, acquire the opacity of enamel.

With salt of phosphorus, fuses, except a silica skeleton, into an almost colourless glass, which becomes opaline on cooling.

With soda fuses slowly into a red grey scoria.

On platina foil the effect of manganese is produced.

(*b.*) *Gadolinite*, from Ytterby, Broddbo, and Finbo, vitreous gadolinite.[2]

---

[1] The specimen used in these experiments was of the purest species, with a yellow granular fracture. The other species often contain a nucleus of vitreous gadolinite, and when we assay this natural mixture by the blowpipe, we obtain complicated effects, from the blending together of the re-actions of the two species. B.

[2] By the formula, calculated from his own analysis, Berzelius considers its composition as, 4 atoms of silicate of yttria +1 atom of sub-silicate of cerium + 1 atom of sub-silicate of protoxide of iron. Klaproth's analysis of gadolinite gave

These gadolinites are of two species, one (*a*) has a perfect resemblance to a piece of black glass; the other (*b*) has a more splintery, and a narrower conchoidal fracture.[1]

(*a*.) *Alone, in the matrass,* no change; gives off no moisture. If we heat the matrass till it begins to fuse, at a certain moment the assay suddenly shines as if it had taken fire; at the same time it dilates a little, and if it be of larger dimensions than usual, it splits here and there, and becomes clear green-grey. Nothing volatile is given off.

On charcoal the same phenomenon occurs; it does not fuse, but the thinnest parts blacken in a strong heat.

(*b*.) *Alone,* intumesces, throws out cauliflower-like ramifications, and becomes white, giving off moisture. It rarely presents an appearance somewhat similar to the species of ignition mentioned above. With respect to the fluxes, both gadolinites behave alike.

With borax they fuse readily into a dark glass, strongly coloured by iron, which in the reducing flame becomes deep bottle green.

With salt of phosphorus fuse with extreme difficulty. The glass assumes an iron tint, and the fragment becomes rounded on the edges, but re-

---

5·5 per cent of glucina, and makes no mention of cerium. C.

[1] The most vitreous of the two contains no trace of glucine. It probably was from the one which breaks into splinters, that Ekeberg obtained 4 per cent of that earth. B.

mains white and opaque, so that the phosphoric acid does not in this instance effect the separation of the silica; it is principally by this character that the gadolinites, *a* and *b*, are distinguished from that from Korarf.

With soda the variety, *a*, changes into a semifused brown-red scoria. The variety, *b*, fuses into a globule, if the quantity of flux be not too great. *On platina foil* neither of them gives the least indication of manganese.

4. *Glucinium.*

1. *Emerald, Beryl.*[1]

Alone, no change with a gentle heat. A thin scale, with a long continued strong heat, becomes rounded on the edges, and forms a blebby colourless scoria. The transparent species turns milk white where the heat was strongest.

With borax fuses into a transparent colourless glass. The chrome green emerald gives a glass, which on cooling assumes a light green tinge, pure and pleasant to the eye.

With salt of phosphorus dissolves slowly, without any siliceous residuum. The assay does not change its appearance, but continually diminishes in size, and at last fuses into a globule, which be-

---

[1] By the formula, calculated from his own analysis, Berzelius states it to be composed of 1 atom of quadri-silicate of glucina + 2 atoms of bi-silicate of alumina. C.

comes opaline on cooling. The chrome green emerald gives a green glass.

With soda fuses into a transparent colourless glass. The yellowish emerald, with a granular fracture from Broddbo, and Finbo, gives in the reducing experiment evident traces of tin.

With solution of cobalt gives an impure very slightly bluish colour.

2. *Euclase.*[1]

Alone, not changed by slight ignition. In a stronger heat, intumesces, whitens and throws out ramifications; at last, in a very intense heat, it fuses on the edges into a white enamel.

With borax intumesces, with slight effervescence, and whitens; then fuses slowly into a transparent colourless glass; the fusion of the last portions is difficult. The glass does not become opaque by flaming.

With salt of phosphorus decomposes, with momentary effervescence; gives a silica skeleton of more than ordinary whiteness, and experiences no further decomposition. The glass is transparent and colourless, but becomes opaline on cooling.

With a little soda fuses into an opaque globule, and, with a larger quantity, into a transparent glass, which becomes opaque on cooling; a still larger portion of soda penetrates into the charcoal: what remains fuses as before. In the reducing experiment we obtain traces of tin.

[1] One atom of silicate of glucina + 2 atoms of silicate of alumina, calculated from the author's own analysis. C.

5. *Magnesium.*

1. *Bitter salt*, from Catalayud in Spain.[1]

Alone, in the matrass, gives off a large quantity of water, free from acidity; the salt fuses, and experiences no further change, at the temperature at which glass melts. When a drop of water is let fall on it, it gives out heat and hardens. If we heat the desiccated salt on charcoal, or in the forceps, it fuses afresh. On charcoal, at a certain temperature, the heat penetrates it almost instantly, and it emits a brilliant light, which lasts as long as we continue the blast. After this, the assay is infusible, having lost its sulphuric acid. Placed on moistened brazil wood paper, or on reddened litmus paper, it tinges both blue.

With borax, and salt of phosphorus, bitter salt behaves like magnesia. (p. 81.)

With soda intumesces, but does not fuse; when moistened the assay exhales a hepatic odour. If we place a particle of the desiccated salt on glass, with an excess of soda, the latter becomes liver coloured.

With solution of cobalt it gives a rose colour of a fine but pale tint.

2. *Magnesite*, from Baudissero.[2]

---

[1] Composed of 1 atom of sulphate of magnesia 60+7 atoms of water, 63=123. C.

[2] Magnesite from Stiria, analysed by Klaproth, gave, mag-

Alone, in the matrass, gives off no water, or only a very small quantity.

On charcoal it cracks a little, and shrinks considerably, after which it acts on moistened brazil wood paper like an alcali.

With borax, salt of phosphorus, soda, and solution of cobalt, it presents the phenomena described under the head Magnesia. (p. 81.)

3. *Boracite,* from Lunebourg.[1]

Alone, in the matrass, no change.

On charcoal fuses and intumesces. It is difficult to obtain the globule transparent, whose surface is bristled over, on cooling, with needle crystals. The glass is yellowish whilst hot, but becomes white and opaque on cooling.

With borax fuses easily into a transparent glass, tinged by iron.

With salt of phosphorus fuses readily into a transparent glass, capable of becoming opaque by flaming. With a larger proportion of boracite, it becomes opaque spontaneously on cooling.

Soda dissolves it. If we only add the quantity of soda necessary to obtain a transparent glass with the assay whilst liquid, on cooling, it forms cry-

---

nesia 48, carbonic acid 49, water 3. *Phillips.* By the formula, it is an anhydrous carbonate of magnesia $(20+22) = 42$. C.

[1] By Vauquelin's analysis, it contains boracic acid 83·4, magnesia 16·6. *Phillips.* By the formula, calculated from Stromeyer's analysis, it is a bi-borate of magnesia $(20+46) = 66$. C.

stals with broad facets, as perfect as those of phosphate of lead. With a larger proportion of soda we obtain a transparent glass, incapable of crystallizing, which is merely a solution of magnesia in glass of borax.

Remark.—If boracite, previously decomposed by soda on charcoal, be pulverised and dissolved in muriatic acid, a slip of paper dipt into the solution, dried and then moistened with alcohol, and burnt whilst moist, tinges the flame green towards the end of the combustion.

4. *Chondrodite*, from Pargas, Åker, and America. The *Brucite* of the Americans.'

Alone, in the matrass, blackens by heat, but gives no appreciable quantity of water; the black colour disappears by roasting in the open air.

On charcoal it is infusible. The most ferruginous chondrodite becomes opaque and brown, on the points that are most strongly heated. The variety containing less iron, that from Åker, for instance, becomes milky white, by the effect of heat.

With borax fuses slowly, but completely, into a transparent glass, slightly tinged by iron. If the glass be saturated with the chondrodite, it loses its transparency by flaming; it does not, however, be-

---

'' *Silicate of magnesia.*—Almost all the magnesian silicates contain a combustible substance, which becomes charred when heated in close vessels. If the mineral be then heated in the open air, the charcoal burns away, and the black colour disappears. Steatite is a remarkable instance. B.

come milky white, but semi-translucid, and crystalline.

With salt of phosphorus decomposes pretty easily, and leaves a semi-transparent siliceous residuum; the glass is clear and colourless, but becomes opaline on cooling.

A small quantity of soda transforms it into a difficultly fusible grey scoria; with a larger quantity it intumesces and becomes infusible.

With solution of cobalt, in a strong heat, it gives a pale red colour, not pleasing to the eye. Chondrodite from Pargas gives a brown-grey, the action of the iron preventing that of the oxide of cobalt on the magnesia.

5. *Pyrallolite*, from Pargas.[1]

Alone in the matrass, gives off water, blackens, and an empyreumatic gas is disengaged. Roasted in the open air it becomes white, swells up, and semi-fuses on the edges into a white, slightly blebby enamel.

With borax fuses readily into a transparent glass.

With salt of phosphorus decomposes, and leaves

---

[1] *Bisilicate of magnesia.*—Nordensköld (*Bidrag till närmare kännedom af Findlands Mineralier och Geognosie*, 1 H. p. 31), considers this mineral as formed of 1 atom of bisilicate of alumina +1 of quadrisilicate of lime +6 of bisilicate of magnesia +2 of water; but its marked resemblance to the Steatite from Bayreuth leads to the idea that the lime and alumina are not essential constituents of it. B.

a semi-transparent silica skeleton. The glass is diaphanous and colourless, and becomes opaline on cooling.

With soda fuses into a transparent glass, slightly coloured by iron.

With solution of cobalt we obtain no very decided colour, till the assay is fused on the edges, when it gives a blue glass, slightly inclining to red.[1]

6. *Ecume de mer*, from Turky, and Valecas in Spain.[2]

Alone, in the matrass, gives off water, exhales an empyreumatic odour, and blackens.

On charcoal recovers its whiteness, contracts considerably, and fuses on the thinnest edges into a white enamel.

With borax and salt of phosphorus behaves like the preceding mineral.

With a sufficient quantity of soda fuses into a transparent glass; too much or too little of the flux makes the glass opaque.

With solution of cobalt assumes a fine lilac colour.

---

[1] In consequence of the large quantity of magnesia contained in this mineral, we should have expected to obtain a red colour; but other minerals equally loaded with magnesia, as the malacolite from Tjôtten, also give a blue glass, whilst some that are less so give a red glass. I do not yet know the cause of this anomaly. B.

[2] By the formula, a trisilicate of magnesia, with 5 atoms of water. C.

7. *Noble serpentine*, from Skyttgrufva, near Fahlun.[1]

Alone, in the matrass, gives off water and blackens.

On charcoal, whitens by heat, and may be fused in a good flame into an enamel, on the thin edges.

With borax fuses slowly into a greenish transparent glass.

With salt of phosphorus behaves like the preceding minerals.

With a certain quantity of soda we obtain, not without difficulty, a semi-liquified mass, resembling enamel. If the quantity of soda be increased, the assay swells up and becomes infusible.

Solution of cobalt gives it a red colour.

8. *Common serpentine*, yellow, translucid from Sala, and Bayreuth.

Alone, in the matrass, and *on charcoal*, behaves like the preceding.

With borax fuses slowly, though with a large quantity, into a transparent glass, incapable of becoming opaque by flaming.

With salt of phosphorus, soda, and *solution of cobalt*, it behaves like the preceding.

9. Seifenstein, from Cornwall.[2]

Alone, in the matrass, gives off water and blackens.

---

[1] Composed, by the formula, of 1 atom of hydrate of magnesia +1 atom of bisilicate of magnesia. C.

[2] Bisilicate of magnesia, with bisilicate of alumina, an atom of each, and 2 atoms of water. C.

On charcoal becomes white again, and afterwards fuses into a colourless glass, full of bubbles.

With borax fuses slowly, but completely, into a transparent glass.

With salt of phosphorus behaves like pyrallolite.

With soda, in a strong heat, may be imperfectly fused into a semi-transparent glass, which does not become more fusible by an additional quantity of soda.

Solution of cobalt gives an impure dark violet colour; the fused edges are blue.

10. *Nephrite, Jade*, from the environs of Geneva.

Alone, in the matrass, scarcely gives off any water, nor blackens.

On charcoal fuses with difficulty into a white glass.

With borax, salt of phosphorus, and soda, behaves like the preceding mineral.

With solution of cobalt gives a black glass in the fused parts.

11. *Hard Fahlunite*, from Fahlun.[1]

Alone, in the matrass, gives off water, loses its colour, and becomes white and semi-transparent.

On charcoal fuses into a colourless, semi-transparent glass.

With borax fuses slowly, but in large quantity, into a diaphanous glass, which does not become opaque by flaming.

[1] One atom of silicate of magnesia + 3 atoms of silicate of alumina + + water. Hisinger's analysis, Afh. i Fysik, &c., B. vi. p. 347, gives this composition, and not that of 1 atom of bisilicate of magnesia + 2 atoms of silicate of alumina, as we find in the place referred to. B.

With salt of phosphorus dissolves like the preceding minerals.

With a certain quantity of soda dissolves into a colourless glass, semi-translucid, and very hard to fuse. A larger quantity of soda causes the glass to intumesce, and renders it infusible.

With cobalt the colour is uncertain till the moment of fusing, when we obtain a blue glass.

12. *Dichröite* (Steinheilit, Cordierite, Watersaphire), from Orrijerfvi, and Sala.[1]

Alone, in a low heat, no change; in a strong heat fuses slowly on the edges into a glass free from bubbles, retaining the original colour and transparence of the stone.

With borax and salt of phosphorus behaves like the preceding.

With soda, no solution; a small dose of this flux occasions a dark grey glassy scoria; with a larger quantity, the assay intumesces, and becomes infusible.

With solution of cobalt the assay becomes black, and the fused edges blue grey.

13. *Peridot.* Olivine, from Auvergne.[2]

---

[1] *Iolite.* According to Gmelin's analysis, it contains, silica 42.6, alumina 34.4, lime 1.7, magnesia 5.8, oxide of iron 1.5, oxide of manganese 1.7. *Phillips.* By the formula it is composed of 1 atom of bisilicate of magnesia + 4 atoms of silicate of alumina. C.

[2] *Chrysolite.* It contains, according to Klaproth, silica 38, magnesia 43.5, iron 19. *Jameson.* By the formula, Berzelius considers it as composed of 1 atom of protosilicate of iron + 4 atoms of silicate of magnesia. C.

Alone, gives off no moisture; turns rather brown, chiefly on the edges, but does not fuse, and preserves its transparence and colour.

With borax, and salt of phosphorus, behaves like the preceding minerals; the glasses are coloured by iron, and give no indication of manganese with saltpetre.

With soda is converted by laborious fusion into a brown scoria.

14. *Chlorite,* from Fahlun.

Alone, in the matrass, gives off water, and, at the fusing point of glass, fluoric acid, which turns brazil wood paper yellow, and leaves a deposit of silica on the glass.

On charcoal fuses into a black globule, with a dull surface.

With borax fuses easily into a dark green glass.

With salt of phosphorus decomposes, and leaves silica. The colour of the glass denotes a considerable quantity of iron.

With soda neither dissolves nor intumesces, but the edges become rounded. *On platina,* gives no trace of manganese.

15. *Diallage.*[1]

[1] I do not know the localities of the two specimens which were the subjects of this and the following experiment; but, as I received them both from M. Hauy, I cannot question the accuracy of their denominations. B.

By referring to the "Nouveau Système," in which I find the same mineralogical formula as in the present work, this must be the *Bronzite* of Jameson, and *Diallage metalloide* of Hauy, which consists, according to Klaproth (on whose ana-

Alone, in the matrass, gives off water not at all acid, crackles, and its colour becomes clearer.

On charcoal fuses slowly on the edge into a greyish scoria.

With borax is converted, by laborious fusion, into a transparent glass, slightly coloured by iron.

With salt of phosphorus decomposes, and gives a siliceous residuum.

With a certain quantity of soda fuses into a greenish grey opaque globule. With a larger dose the assay intumesces and becomes infusible. On platina foil it gives no indication of the presence of manganese.

16. *Hyperstene.*[1]

Alone, in the matrass, crackles slightly, and gives off water free from acidity, but its appearance does not sensibly alter.

On charcoal fuses easily into a greyish green opaque glass.

With borax fuses easily into a greenish glass.

With salt of phosphorus no sensible decomposition; the assay becomes rounded on the edges, and fuses extremely slowly.

---

lysis the formula is calculated), of silica 60, magnesia 27·5, iron 10·5, water, 0·5. *Jameson.* Berzelius considers its composition to be, 1 atom of bisilicate of protoxide of iron + 3 atoms of bisilicate of magnesia. C.

[1] By Klaproth's analysis it contains, silica 54·25, magnesia 14, alumina 2·25, lime 1·5, oxide of iron 24·5, water 1. *Jameson.* By the formula, calculated on the preceding analysis, it consists of 1 atom of bisilicate of protoxide of iron +1 atom of bisilicate of magnesia. C.

With soda behaves like the preceding mineral.

17. *Sordawalite*, from Sordawala in Finland. It appears to be a mixture of 1 atom of phosphate of magnesia + 2 atoms of water, with a stony fossile, formed of 1 atom of bisilicate of magnesia + 2 atoms of bisilicate of protoxide of iron + 3 atoms of bisilicate of alumina.[1]

Alone, in the matrass, gives off a great deal of water, free from acidity.

On charcoal fuses, without intumescence, into a black globule, which becomes grey, and assumes a metallic lustre in the reducing flame.

With borax fuses easily into a glass tinged green by iron.

With salt of phosphorus decomposes readily, and leaves a silica skeleton.

With a small quantity of *soda* fuses into a black globule; with a larger dose, intumesces, and is converted into a distorted (*anfractueuse*) scoria. On platina foil it gives indications of manganese.

With boracic acid and iron I have not been able to obtain from it a phosphuret of iron.

18. *Magnesian garnets.*—Those from Syria, Orrijerfvi, Hollandsos, &c., behave like almandine, and the magnesia they contain cannot be detected by the blowpipe.

19. *Spinel*, from Ceylon and Åker.[2]

[1] According to Nordensköld's analysis, 1 H, p. 86, et seq. of the work quoted above. B.

[2] Vauquelin's analysis of spinel, or balass ruby, gave, alumina 82·47, magnesia 8·78, chromic acid 6·18, loss 2·57.

Alone, no change.[1] The red spinel from Ceylon indeed turns brown, and even blackens and becomes opaque before the blowpipe, but, on cooling, it recovers its colour in the following manner—seen by transmitted day-light, it first displays a fine chrome green, then becomes almost colourless, and, lastly, resumes its ruby tint.

With borax it fuses slowly into a transparent glass, with little colour. The spinel from Åker, sometimes contains lime in its interstices, in which case it fuses with effervescence into a glass that may be made opaque by flaming.

With salt of phosphorus fuses with difficulty in fragment, but readily and without residuum when pulverised. The glass generally presents the characteristic colours of iron, but that of the Ceylon spinel assumes, after cooling, a sensible, though faint tint of chrome green. It does not become opaline.

With soda does not fuse, but swells up; on platina foil gives slight traces of manganese.

20. *Pleonaste*, from Ceylon and Somma.[2]

Jameson. Berzelius considers it, from his own analysis, to be a sexaluminate of magnesia. C.

[1] Before the gas blowpipe it fused readily with partial combustion and loss of weight. One solid angle of an octohedral crystal was entirely burnt off and volatilized. Gas Blowpipe, p. 58. C.

[2] One atom of aluminate of protoxide of iron + 1 atom of sexaluminate of magnesia; calculated from the analysis of Collet Descotils (Journal des Mines, xxx. 421). The chemical composition of this mineral requires fresh examination. B.

Alone, suffers no change, except that in a very strong heat, it assumes the blue colour of the glassy scoriæ of the graduated furnaces (fourneaux gradués). It does not fuse, even in powder, but presents a vitreous appearance on the edges.

With borax fuses into a transparent glass, of a dark green colour, precisely like that given by iron.

Salt of phosphorus scarcely has any action on it in fragment; but, when pulverised, it fuses readily with this flux into a transparent glass coloured by iron, and leaves no residuum.

With soda intumesces, and gives a black scoria, which does not become fusible by a larger quantity of the flux.

21. *Hydrate of magnesia*, from New Jersey.[1]

Alone, in the matrass, gives off water; both before and after ignition, it restores the blue colour of reddened litmus paper.

On charcoal thickens (s'épaissit) in the longitudinal direction of the laminæ, crackles a little and becomes milk white, but does not fuse.[2]

With the fluxes, and solution of cobalt, behaves like pure magnesia.

6. *Calcium.*

1. *Sulphate of lime*, Gypsum.
(*a.*) *Anhydrous gypsum*, Anhydrite.[3]

---

[1] Magnesia and water. C.
[2] See Note, p. 61. C.
[3] Lime and sulphuric acid, without any water. C.

Alone, in the matrass, gives off no moisture, or only traces of it.

In the forceps fuses with difficulty, in the oxidating flame, into a white enamel.

On charcoal, in a good reducing flame, decomposes, and then acts as an alcali on brazil wood paper, and exhales the smell of liver of sulphur when moistened.

With borax fuses with effervescence into a transparent glass, which, when cold, is yellow, or brown-yellow. If the proportion of gypsum be increased, the globule becomes brown and opaque on cooling.

With the other fluxes it behaves like pure lime.

With fluate of lime fuses readily into a transparent globule, which assumes the whiteness of enamel on cooling; by a prolonged blast it intumesces and becomes infusible.

Glass of soda developes the colour of liver of sulphur.

(b). *Common gypsum.*[1]

Alone, in the matrass, gives off water, and becomes milky white. It afterwards behaves like the preceding.

2. *Fluate of lime.*[2]

(a). *Fluor spar.*

Alone, in the matrass, and at a heat much below incipient redness, it often exhibits a greenish light,

---

[1] One atom of sulphate of lime, with 4 atoms of water. C.

[2] Lime and fluoric acid, or rather fluorine and calcium. C.

visible in the dark. In a more intense heat it decrepitates strongly, and gives off very little water.

On charcoal it may be fused by a good heat into an opaque globule.

With borax, and salt of phosphorus, fuses with the greatest facility into a transparent glass, which, at a certain point of saturation, becomes opaque.

With a little soda fuses into a diaphanous glass, which, after a long blast, loses its transparence on congealing; if we increase the quantity of soda, it is transformed into a difficultly fusible enamel, which remains on the charcoal, whilst the excess of soda is absorbed.

With gypsum fuses readily into a transparent glass, which becomes opaque on cooling. (See *Gypsum.*)

(*b*). *Yttrocerite,* from Finbo.[1]

Alone, in the matrass, gives off a little empyreumatic water; the dark coloured variety becomes white.

On charcoal does not fuse by itself; but, by the addition of gypsum, it melts into a globule opaque at all temperatures.

With borax, salt. of phosphorus, and soda, behaves like fluor spar, except that the glasses become yellow in the oxidating flame, and retain that colour

---

[1] Berzelius' analysis gave, fluate of lime 68·18, fluate of yttria 10·6, fluate of cerium 20·22. *Phillips.* The formul represents it as consisting of an atom of fluate of lime, an atom of fluate of yttria, an atom of sesqui-fluate of cerium. C.

whilst hot. They lose their transparency sooner than those of fluor spar.

(c). *Yttrocerite*, from Broddbo.

Decrepitates slightly without fusing, and passes from milk white to brick red (this change of colour is not uniform); does not fuse with gypsum, and, in other respects, behaves like fluate of cerium, which it contains in very considerable quantity.

3. *Carbonate of lime.*

(a). *Calcareous spar.*

Alone, in the matrass, gives off no moisture.

On charcoal becomes caustic by heat, and shines with peculiar brightness as soon as all the carbonic acid is expelled; heats with water, and acts as an alcali on reddened litmus paper. Ferruginous or manganesian calcareous spar blackens by heat.

With the fluxes, in which it dissolves with effervescence, it behaves as stated under the head "Lime," p. 79. Those varieties which contain iron or manganese give a coloured glass.

(b). *Arragonite.*

Alone, in the matrass, no change at first, at a temperature much beyond that of boiling water, but, when nearly red hot, it intumesces, and falls into a coarse, light, white powder; at the same time, a very small quantity of water collects in the neck of the matrass, less even than that afforded by other minerals which contain only water of decrepitation. With the fluxes it behaves like the preceding mineral.

4. *Bitter spar* (magnesian spar).[1]

Not distinguishable by the blowpipe from calcareous spar.

5. *Phosphate of lime.*[2]

(*a*). *Moroxite,* from Arendal and Pargas.

Alone, unalterable in a solid shaped fragment, but in spangle it fuses on the edge, with a very strong heat, into a colourless, translucid glass; it is one of the most difficult minerals to fuse.[3]

With borax fuses slowly into a transparent glass, which turns milk white by *flaming,* and, with a large proportion of moroxite, becomes opaque on cooling.

With salt of phosphorus fuses in large quantity into a transparent glass, which, when nearly saturated, becomes opaque by cooling, and exhibits facets less distinct than those which result from the crystallization of phosphate of lead. When perfectly saturated, it congeals, without regularly crystallizing, into a milk white globule.

With soda swells up and effervesces; the soda penetrates the charcoal, and leaves a white mass on the surface.

---

[1] One atom of carbonate of lime + 1 atom of carbonate of magnesia. C.

[2] *Apatite.* The *asparagus stone* from Spain, analysed by Klaproth, gave, lime 54·28, phosphoric acid 55·72. *Phillips.* Hence, it consists of 1 atom of lime 28 + 1 atom of phosphoric acid 28 = 56. Moroxite is another variety of apatite. C.

[3] Before the gas blowpipe crystallized *apatite* fused into a *black shining slag. Compact apatite* gave " a *white enamel,* resembling, as to external appearance, *spermacetti.*" Gas Blowpipe, p. 52.

With boracic acid fuses with extreme difficulty, but gives, *with metallic iron,* a regulus of phosphuret of iron.

(*b*). *Radiated phosphate of lime, phosphorite,* from Estremadura.

In the matrass, gives off a little water, fuses rather more readily than the preceding, into a white enamel. In other respects it behaves like moroxite.

(*c*). *Phosphate of lime,* mammoth's teeth, dug up at Kannstadt.

In the matrass contracts considerably, and gives off a large quantity of water.

On charcoal blackens at the extreme part acted on by the flame; does not fuse, but becomes rounded and semi-transparent at the same part. In other respects, exhibits the same effects as the preceding minerals.

6. *Datholite,* and

7. *Botryolite,*[1] both from Arendal. They behave alike, and as follows:—

Alone, in the matrass, they give off a little water.

On charcoal they intumesce a little, like borax,

---

[1] Klaproth's analysis of datholite gave, silica 36·5, lime 35·5, boracic acid 24, water 4; his analysis of Botryolite gave, silica 36, lime 39·5, boracic acid 13·5, water 6·5, oxide of iron 1. *Jameson.* From these analyses, Berzelius calculates his formulæ, which give for the former, 1 atom of bi-borate of lime + 1 atom of tri-silicate of lime (according to the tables in the essay on chemical proportions) + 1 atom of water; for the second, 1 atom of borate of lime + 1 atom of trisilicate of lime + 1 atom of water. The weight of an atom of lime is 28, that of boracic acid 23, and of silica 16. C.

and fuse into a transparent glass, which is either colourless, pale rose, or iron green, according to the purity or colour of the assay.

With borax fuse readily into a transparent glass, whose colour varies like the preceding.

With salt of phosphorus dissolve, except a silica skeleton. If we add a fresh portion of the assay, the glass loses its transparence, and, lastly, becomes enamel white.

With a little soda fuse into a transparent glass. With a larger dose the glass becomes opaque on cooling, and, with a still larger, the whole mass penetrates the charcoal.

With gypsum they fuse, but with more difficulty than fluor spar, into a diaphanous globule, which retains its transparence on cooling.

With solution of cobalt, an opaque blue glass.

Remark.—If we pulverise either of these minerals, moisten the powder with a drop of muriatic acid, and suffer it to dry on a slip of thin paper; on wetting the paper with alcohol, and setting it on fire, the flame towards the end of the combustion will be tinged green. This phenomenon does not occur with boracite, unless it be previously ignited with soda.

8. *Arseniate of lime*, pharmacolite.[1]

Alone, in the matrass, gives off much water, free from acidity; no arsenious acid sublimes; the

[1] Klaproth's analysis gave, lime 25, arsenic acid 50·54, water 24·46. *Jameson.* By the formula, it contains 1 atom of acid + 1 of base + 6 of water. C.

assay loses its transparence, and resembles an effloresced salt, but retains its form.

In the forceps fuses in the exterior flame into a white enamel.

On charcoal, in the interior flame, fuses more readily, giving off the smell of arsenic, into a semi-translucid globule, sometimes inclining to blue, that is, if the assay, as generally happens, be mixed with arseniate of cobalt.

With borax, and salt of phosphorus, behaves like lime, or the salts of lime with volatile acids in general, but gives off, in dissolving, abundant fumes of arsenic.

With soda decomposes with great disengagement of arsenic. The lime remains on the charcoal.

9. *Tungstate of lime.*[1]

Alone, in the matrass, no change.

On charcoal, in a strong heat, the thin parts fuse into a semi-transparent glass.

With borax fuses easily into a transparent glass, which soon becomes opaque, milk white, crystalline, and incapable of being coloured by the reducing flame, even with tin.

With salt of phosphorus fuses readily in the exterior flame into a colourless transparent glass; in the interior flame, the glass assumes a green colour, which it retains whilst hot, but which turns to a

---

[1] A specimen lately analysed by Berzelius consisted of tungstic acid 80·417, lime 19·460. *Phillips.* It contains by the formula, 1 atom of base + 1 of acid. The weight of an atom of tungstic acid is 120. C.

fine blue on cooling. If we add tin, the glass takes a darker tint and becomes green. After a long blast, with a sufficient quantity of tin, tungsten precipitates, and the colour changes at last to a very pale greenish yellow.

With soda forms a white intumescent scoria rounded on the edges.

10. *Uranate of lime,* uranite; the yellow variety from Autun, the green from Cornwall.[1]

Alone, in the matrass, gives off water, and becomes straw yellow and opaque.

On charcoal intumesces slightly, and fuses into a black globule, with a semi-crystalline surface.

With borax and salt of phosphorus fuses readily into a transparent glass, of a dark green colour in the oxidating flame, and fine green in the reducing. With tin, the glass formed with the green variety becomes red and opaque, from the oxidation of the copper which it contains.

With soda does not fuse, but gives a yellow scoria. The Autun uranite gives no metallic particles in the reducing experiment, but the green uranite from Cornwall exhales the smell of arsenic, and gives white metallic grains, which are nothing else but an alloy of arsenic and copper, derived from the arseniate of copper by which the mineral is coloured.

---

[1] One atom of base + 1 of acid + 6 of water. C.

11. *Sphene,* titane silicéo-calcaire: formula of its composition not ascertained.

Alone, in the matrass, gives off a small quantity of apparently merely hygrometric water. The yellow sphene suffers no change. The brown variety becomes yellow, but retains nearly its original transparence. A variety from Frugord in Finland, presents during this change of colour a phenomenon of ignition, similar to that described under the head of vitreous gadolinite, but much less intense.

On charcoal, or *in the forceps,* fuses on the edges, with slight intumescence, into a dark coloured glass. The unfused portion retains its clear yellow colour and semi-transparence.

With borax fuses pretty readily into a clear yellow transparent glass, which becomes brown by an additional quantity of sphene, but cannot be made to develope the characteristic colour of titanium by the reducing flame.

With salt of phosphorus dissolves with difficulty; the unfused portion becomes milk white. In a good reducing flame the characteristic colour of titanium is developed, and, by the addition of tin, with the utmost facility. After a long blast the glass becomes opaline on cooling.

With soda fuses into an opaque glass, which no proportion of soda can make transparent. When cold it is similar to that formed with pure oxide of titanium. With a large quantity of soda the glass is absorbed by the charcoal, and usually gives a little iron in the reducing operation.

12. *Tabular spar*, from Nagyag, Perhoniemi, Pargas, Gökum, and Capo di Bovi.[1]

Alone, in the matrass, no change.

On charcoal melts on the edge into a semi-transparent colourless glass globule. Requires a very strong heat for its perfect fusion, and bubbles up a little at intervals.

With borax fuses easily, and in large quantity, into a transparent glass, which cannot be made opaque by continuing the blast.

With salt of phosphorus decomposes, and leaves a silica skeleton, full of flaws (glaceux). The glass becomes opaline on cooling.

With a little soda fuses into an enamel white, blebby glass. A large quantity makes the assay infusible and intumescent.

With solution of cobalt it is much more difficult to fuse than by itself, or without addition; the fused edge is blue.

---

[1] This mineral, formerly only found in Transylvania, has lately been met with in the places above named, and appears pretty frequently to accompany primitive limestone. It has been analysed by Laugier, Bonsdorff, and Rose, and the experiments with the blowpipe confirm its identity with the specimens from those places. It seems to have been generally considered as a white variety of amphibole or tremolite. B.—Klaproth's analysis of tabular spar gave silica 50, lime 45, water 5. *Jameson*. By the mineralogical formula it is a bisilicate of lime. C.

13. *Amphibole*, hornblende, tremolite, asbestos, actinote, &c.

In a system of mineralogy founded on crystallization, under the same name, indicating a certain crystallographical form, are comprehended a certain number of minerals, which, though they differ in colour, and in many other respects, have nevertheless one common figure: but it is impossible to deduce an equivalent generic character from the phenomena which these minerals present before the blowpipe; of which amphibole, pyroxene, and garnet, may be taken as instances.

Mitscherlich has proved that certain bases, saturated with the same acid to the same degree, affect the same crystalline form, and he has particularly shown that lime, magnesia, and the protoxides of iron and manganese compose in this way a class of *isomorphous* bases. He has demonstrated further, by experiments made in the moist way, that isomorphous salts have the property of crystallizing *in common* (*en commun*), concurring in an uniform manner in the structure of one and the same crystal, without being mutually connected by any chemical affinity, and consequently without being confined to definite proportions. Now it seems that the same operation has occurred in nature, when minerals, in crystallizing, have separated from one another. If this inference be correct, we can comprehend the hitherto enigmatical fact, of the absolute identity of geometrical form in minerals, whose analyses present very remarkable differences. Hence, in order to judge with certainty as to the results of his analyses, the chemist need only ascertain which are isomorphous combinations. From what we have seen, the silicates of lime and magnesia, and those of the protoxides of iron and manganese, may meet in the same crystal in the same degree of saturation, and their relative quantities may vary, although the form of the crystal remain the same. Whence it follows that crystals of amphibole may not only exhibit very different colours, but also very different phenomena when acted on by the blowpipe. Although these theoretical ideas suppose an indefinite number of possible mixtures, amongst these the most common of the silicates, (which experiment sufficiently confirms), yet, when we are better acquainted with them, a few general phenomena will suffice to ascertain the nature of the constituent parts of a mineral crystallized after the manner of amphibole.

At present I can only give the characteristic phenomena of the most significant species of those which affect this crystalline form. According to M. Bonsdorff's careful and accurate analysis of some of the purest species, they appear to be composed of one volume[1] of trisilicate of lime, with three volumes of bisilicate of magnesia. All these were nearly or quite colourless. But we have another kind of hornblendes, which, though they affect the most regular form of the amphibole crystals, differ considerably from them in chemical compo-

[1] Atom. C.

sition. They not only contain a new base entirely heteromorphous with respect to the before-mentioned species, namely alumina, but in many of them, the silica is not in sufficient quantity to form silicates; in other words, the oxygen of the silica is often in less quantity than that of the bases, and the more alumina there is in the mineral, the less silica it contains. To be satisfied of the accuracy of this, it is enough to calculate the analyses of some black species of hornblende, published by Hisinger, in the Journal entitled *Afhandl. i Fysik, Kemi*, &c, vi. 199; as well as the analyses recorded by Klaproth in his work, under the article *Hornblende*. The question, therefore, is, whether the constitution of these black hornblendes be absolutely new. Now, if we may hazard a conjecture on the subject, it is most probable, that these species contain, besides bisilicates and trisilicates of magnesia, lime and protoxide of iron, an aluminate resulting from the combination of the alumina with the excess of the bases just mentioned, which aluminate may be conceived to be isomorphous with one of the silicates that accompany it. I have thought it right to interrupt the detail of the phenomena which minerals present in pyrognostic assays, by this little digression, in order that the reader may more easily judge of the results of these experiments. In general, minerals of the hornblende species fuse more readily than the pyroxenes, of which we shall speak immediately afterwards, and which come near them in composition. This seems to be owing, on one hand, to their containing more than one atom of magnesia to an atom of lime, and on the other, to their being rich in bisilicate of iron,—an easily fusible substance. However, we meet with species of amphibole which fuse with difficulty, and very ferruginous pyroxenes whose fusion is easily effected. I shall divide the amphiboles into two classes; the first, *a*, will include those which contain no alumina; the second, *b*, the generally black aluminous amphiboles.

(*a.*) *Amphiboles not containing alumina.*

(α.) *Colourless amphibole,* from Gullsjö in Wermlande.[1]

Alone, in the matrass, no change, only a little hygroscopic water is given off.

In the forceps fuses readily, with slight bubbling, into a semi-transparent glass; the part next the fused mass becomes milk white. Every time that we repeat the fusion of the glass it begins by bubbling, which then subsides again.

With borax fuses slowly into a colourless transparent glass.

With salt of phosphorus not decomposed; the

[1] In this, and most of the following minerals, M. Bonsdorff found some fluoric acid, which he considers as forming a neutral combination in those fossils, with a corresponding quantity of lime not expressed in the formulæ. B.

The formulæ are calculated from Bonsdorff's analyses; that of the mineral under consideration gives its composition as 1 atom of trisilicate of lime + 3 atoms of bisilicate of magnesia. C.

assay remains milk white throughout, and is rounded on the edges; after a long blast the glass becomes opaline on cooling.'

With a very small quantity of soda fuses into a transparent glass. A larger quantity of the flux causes it to intumesce, and converts it into a white infusible scoria.

Solution of cobalt developes a rose colour in the fused parts.

(β.) *Grammatite*, from Fahlun.[2]

Alone, not changed by gentle ignition; in a strong heat it intumesces a little, cracks longitudinally, and turns milk white; lastly, in a very strong heat fuses with effervescence into a distorted, almost opaque grey white mass.

With borax and salt of phosphorus behaves like the preceding mineral.

With soda, in suitable quantity, fuses into a transparent glass; with too large or too small a dose the glass becomes opaque when cold.

With solution of cobalt gives a dark red in the fused part, and a fine bright red all around it.

(γ.) *Asbestiform Tremolite*, from Sheffield, in Massachusets.

---

[1] This property, with a few exceptions, is common to all the amphiboles. There is, however, a pretty good method of decomposing them by salt of phosphorus; it consists in fusing them with a very small quantity of the flux, with which we merely cover their surface. When the assay begins to puff up, the phosphoric acid penetrates it; it then intumesces, decomposes, and becomes full of flaws (glaceuse).

[2] Same composition as colourless amphibole. C.

Alone, bubbles and fuses with great difficulty into a vitreous mass, generally presenting a delicate radiated crystallization on the surface.

With borax and salt of phosphorus behaves like the preceding minerals.

With soda fuses very readily into a transparent glass, which requires a large quantity of soda to render it opaque.

With solution of cobalt the fused edges are coloured red.

(δ.) *Asbestus*, from the Tarentaise.[1]

Alone, fuses very readily into a slightly greyish globule, similar to enamel; its surface is not vitreous.

With the fluxes behaves like α.

(ε.) *Asbestiform actinote*, from Taberg near Philipstad,[2] and from Fahlun.

Not sensibly altered by moderate ignition. In a brighter heat it whitens, and then fuses with slight bubbling into a yellowish brown opaque glass.

With borax fuses readily into a glass slightly coloured by iron.

Salt of phosphorus has not a very powerful action on it. The crystalline radii do not change

---

[1] In Savoy. Chenevix's analysis gave, silica 59, alumina 3, lime 9, magnesia 20. *Phillips.* By the formula it consists of 1 atom of trisilicate of lime + 3 compound atoms of bisilicate of magnesia and protoxide of iron. C.

[2] One atom of trisilicate of lime + 3 compound atoms of bisilicate of magnesia and protoxide of iron. C.

their appearance, and the glass becomes slightly opaline on cooling.

With a certain proportion of soda it gives an opaque greenish glass, and with a larger quantity an infusible intumescent mass.

With solution of cobalt fuses and reddens on the edges.

Remark.—The intensely green actinotes give a glass more decidedly tinged by iron; in other respects they behave like ε.

(ζ.) *Byssolite*, from Oisans.[1]

Alone, fuses into a brilliant black globule.

With borax, fuses readily into a glass coloured by iron.

With salt of phosphorus dissolves with difficulty. The crystalline radii which first present themselves undergo complete solution; the rest remain untouched.

With soda gives a black glass; a larger quantity converts the assay into a black scoria. On *platina* traces of manganese are perceptible.

(*b.*) *Aluminous amphiboles.*

(η.) *Grammatite*, from a limestone quarry at Åker.[2] Whitens, without splitting, in a strong heat,

---

[1] The author seems doubtful as to its composition. The formula (to which a ? is annexed), represents it as a bisilicate of lime, magnesia, protoxide of manganese and peroxide of iron, an atom of each. C.

[2] This grammatite contains from 4 to 14 per cent. of alumina. B. The formula, calculated from Bonsdorff's analy-

and fuses more easily than the grammatite from Fahlun (nearly like that Gullsjö) into a transparent almost colourless glass.

With borax fuses readily into a transparent glass.

With salt of phosphorus intumesces, and does not wholly decompose, but is converted into a translucid mass, the central part of which retains its hardness.

With soda fuses into an opaque glass, not readily liquifiable; with a larger quantity it at first intumesces, but afterwards in a strong heat fuses into a globule.

With solution of cobalt fuses more difficultly, and only on the edges, into a fine dark blue glass.

(θ.) *Black primitive hornblende,* from Slättmyra.[1]

Alone, scarcely intumesces, and fuses without the least bubbling into a black brilliant globule.

With borax fuses into a glass strongly tinged by iron.

With salt of phosphorus not decomposed. The assay does not change colour, but becomes rounded on the edges. The vitrified flux exhibits a slight tint of iron. In this respect this mineral resembles the non-aluminous amphiboles.

With soda fuses readily into a black brilliant

---

sis, considers it as 1 atom of trisilicate of lime + 3 atoms of bisilicate of magnesia. C.

[1] It contains 7½ per cent. of alumina. (Hisinger, Afh. vi. 201.) B.

glass, which a larger portion of soda renders still more fusible, but its surface then loses its brilliancy and becomes crystalline, and the assay assumes a deep brown colour.

(ι.) *Black Hornblende in large laminæ*, from Taberg, in the metalliferous district of Nora.

Behaves like the preceding.

(κ.) *Dark green lamellar hornblende*, from Annaberg, in Saxony.

Alone, fuses with effervescence and intumescence, into a black brilliant glass.

With borax fuses readily into a glass but little coloured.

Salt of phosphorus decomposes it after the blast has been continued some time, and leaves a silica skeleton. The globule is colourless, and becomes opaline on cooling.

With soda, same phenomena as with η.

(λ.) *Black crystallized hornblende*, from Pargas.[1]

Fuses readily with violent bubbling into a brown grey opaque glass.

With borax fuses easily into a clear greenish glass.

Salt of phosphorus decomposes it readily, and converts it into a flawy mass. The fused glass is tinged by iron whilst hot, and becomes opaline on cooling.

With soda fuses with difficulty into a grey brown glass, which it is not easy to obtain in a globule.

[1] It contains 12 per cent of alumina, according to the analyses of Hisinger and Bonsdorff. B.

(μ.) *Pargasite*, or crystallized clear green, diaphanous hornblende, from Pargas.

Behaves like the preceding, except that its glasses are less coloured. It differs in composition only in containing a smaller proportion of iron.

14. *Pyroxene*.

It is with this mineral as with amphibole, whose figure is common to a pretty large number of different compounds formed by the silicates of the four bases of which we have spoken already; but, what is very interesting, pyroxene is always formed by the bisilicates *of those same bases*, according to the truly curious researches of M. H. Rose, on the composition of the pyroxenes. From the analyses hitherto made, it appears that the pyroxenes contain as many, or more atoms of silicate of lime, as of silicate of magnesia. Colourless malacolite, and the colourless transparent pyroxenes, which Laugier, Hisinger, Bonsdorff, Wachtmeister and Nordenskiöld have analysed, operating on specimens from different places, all gave 1 atom of bisilicate of lime + 1 atom of bisilicate of magnesia; but the clear green opaque species often gave an excess of silicate of lime, and the black usually contain, like certain species of amphibole, more or less alumina, which has no influence on either the form or cleavage of the mineral. In some pyroxenes the bisilicate of magnesia is replaced by bisilicate of protoxide of iron, as in *hedenbergite*, which, according to Rose's experiment, is 1 atom of bisilicate of lime + 1 atom of bisilicate of protoxide of iron, and contains no magnesia. The pyroxenes thus divide into four principal classes, according to the proportions and nature of their constituent parts. The composition of the first class, is 1 atom of bisilicate of lime + 1 atom of bisilicate of magnesia, and comprehends the colourless, transparent pyroxenes; the second includes those which contain more atoms of lime than of magnesia; the third, those in which bisilicate of iron is an essential element, and lastly, the fourth is composed of the aluminous pyroxenes, which are generally black.

(a.) *White or transparent Pyroxene.*

{
Diopside, or *Alalite*, from Piémont.
White malacolite, from the limestone quarry of Tammare, in Finland.
White malacolite, from Tjötten in Norway.
Pale green salite, from Sala.[1]
}

Alone, fuses with bubbling into a colourless, semi-transparent glass.

With borax fuses readily into a diaphanous glass.

With salt of phosphorus decomposes slowly, and leaves a silica skeleton. The glass is transparent, and becomes opaline on cooling. Diopside retains its transparence and aspect a long time, but at last it intumesces and is converted into a silica skeleton.

With a small quantity of soda intumesces, and fuses readily into a transparent glass, which a

[1] All these are bisilicates of lime and magnesia, an atom of each. C.

larger dose of soda renders opaque and less fusible; every time we add a fresh portion of the flux, the glass intumesces before it fuses.

With solution of cobalt the malacolite from Finland fuses, and reddens on the edges. Diopside presents on the fused edges a red colour; when fused into a globule it is violet. The malacolite from Tjötten gives a blue [1] colour inclining to red.

Remark.—Amongst the minerals comprehended under the name of *Salite,* is found at Sala a malacolite, which, in point of aspect, differs nothing from common salite, except in the inferior brilliancy of its surface; its form and colour are the same. It contains, however, only 5 per cent. of lime; the rest is, according to Rose's analysis, a silicate magnesia. It presents the following phenomena before the blowpipe.

Alone, in the matrass, it gives off water.

On charcoal it becomes grey white, but does not fuse either in fragment or in powder, it merely agglutinates on the edges, which become brown and vitreous.

With borax and soda it behaves like the preceding minerals.

With salt of phosphorus it decomposes, with nearly as much difficulty as the preceding minerals, but the assay neither retains its transparence nor colour; it first becomes enamel white, then in-

[1] Can this colour be derived from the $\frac{1}{2}$ per cent. of alumina which this mineral contains? B.

tumesces, slightly, and lastly, is converted into a silica skeleton.

With solution of cobalt it gives an impure red colour both on the fused and unfused parts. The oxide of cobalt increases its fusibility, whereas it produces a contrary effect with the malacolites which contain more lime.

(*b.*) *Pyroxenes containing a greater number of atoms of lime than of magnesia.*

Malacolite, from Björnmyresveden.[1]

Fuses easily into a dark coloured glass.

With borax, salt of phosphorus, and soda, behaves like the preceding minerals, except that its glass is coloured by iron.

(*c.*) *Pyroxenes which contain a bisilicate of protoxide of iron as an essential element.*

(α.) *Hedenbergite*, from Mormorsgrufva, near Tunaberg.[2]

Alone, gives off no water, or only a little hyrometrical moisture, which when the assay is heated to the fusing point of glass, gives, by the test of litmus paper, some traces of acidity.

---

[1] Three atoms of bisilicate of lime + 2 atoms of bisilicate of magnesia + 1 atom of bisilicate of protoxide of iron. C.

[2] One atom of bisilicate of lime + 1 atom of bisilicate of protoxide of iron. C.

This mineral was considered on the authority of Hedenberg, who first analysed it, as a bisilicate of protoxide of iron combined with water, but according to Rose's analysis it is a double bisilicate of lime and protoxide of iron, and presents all the external characters of dark green malacolite. B.

In the forceps it fuses with exceeding slight effervescence into a black shining glass.

With borax fuses easily into a glass strongly coloured by iron.

With salt of phosphorus, decomposes slowly, and gives a silica skeleton, in which the undecomposed nucleus is distinguished by its black colour. The ferruginous colour of the glass disappears on cooling.

With soda fuses readily into a black glass, whose surface becomes dull by a fresh quantity of soda. This mineral absorbs a much larger portion of the flux than any of the preceding before it is converted into a scoria.

(β.) *Dark green pyroxene*, from Taberg, near Philipstad, and from Arendal; *dark red malacolite*, from Degerö in Finland; all behave like the preceding mineral.

(*d*.) *Aluminous pyroxenes*, the greater part of which are black.

(α.) *Pyroxene*, from Pargas.[1]

(β.) *Pyroxene*, from Auvergne, from Lava.

They behave like pyroxenes in general, but are much more difficult to decompose by *salt of phosphorus*, and even almost indecomposable by that flux. The assay changes from its original black, and becomes translucid, and almost colourless.

[1] One compound atom, formed of an atom of silicate of protoxide of iron + 2 of silicate of alumina, combined with three compound atoms, each formed of 1 atom of bisilicate of lime + 1 of bisilicate of magnesia. (Nordenskiöld). C.

With soda they form a more difficultly fusible glass, so that they become infusible with a quantity of soda, with which the essentially ferriferous pyroxenes, of a dark green colour, still vitrify very readily.

15. *Epidote.*

This name also denotes a certain crystalline form, which comprehends minerals of different composition; namely,

(α.) *Zöizite*, from Bareuth, and Kärnthen.[1]

Alone, intumesces and expands transversely to the direction of the laminæ; at the first impression of the flame, numerous small bubbles form, which subside with a stronger heat. Fuses on the outer edges into a slightly yellowish transparent glass, but the intumesced mass becomes afterwards extremely difficult to fuse, and forms a vitreous scoria.

With borax intumesces, and fuses into a diaphanous glass.

With salt of phosphorus intumesces, decomposes readily with effervescence, and gives a siliceous skeleton.

With a very small portion of soda fuses into a slightly greenish glass. The usual quantity of the flux gives an intumescent, white, infusible mass. *On platina foil*, traces of manganese are perceptible.

With solution of cobalt, a blue glass.

---

[1] One atom of silicate of lime + 2 atoms of silicate of alumina. C.

(β). *Pistacite*, from Oisans, Hellestad, Arendal, Taberg, and Orrajerfvi.[1]

Alone, fuses at first in the extreme parts, then intumesces, and is converted into a ramified mass, presenting in miniature a cauliflower appearance of a deep brown colour; in a strong flame it blackens and becomes rounded, but does not completely fuse. The variety from Arendal, which, according to the analyses, is the most ferruginous of all, is also remarkably more fusible than the others.

With borax first intumesces and then fuses into a glass coloured by iron.

With salt of phosphorus decomposes readily with intumescence, and leaves a silica skeleton.

With a little soda fuses laboriously into a dark glass, which a larger dose of soda converts into an infusible scoria.

(γ). *Manganesian epidote*, from Saint-Marcet, in Piémont.

Alone fuses very easily, with bubbling, into a black glass.

With borax fuses with effervescence into a trans-

---

[1] One atom of silicate of lime + 1 atom of silicate of peroxide of iron + 4 atoms of silicate of alumina. C.

Perhaps this formula is correct only for the variety of pistacite from Oisans in France; in general, pistacite is an epidote in which the silicate of protoxide of iron takes the place of a greater or less quantity of silicate of lime. The fine epidote from Oisans differs from zöizite by having an atom of bisilicate of lime replaced by an atom of another isomorphous silicate, namely, bisilicate of protoxide of iron; whence it follows that these two minerals affect the same crystalline form. B.

parent glass, which, in the exterior flame, assumes an amethyst colour; and in the interior, whilst hot, is tinged by iron, but becomes colourless on cooling.

With salt of phosphorus intumesces and decomposes, leaving a silica skeleton. The glass does not assume the colour indicative of manganese, but is tinged by iron whilst hot.

With soda behaves like the preceding minerals.

16. *Prehnite* and *Koupholite.*

According to Klaproth's analysis, these minerals, except the small quantity of water they contain, must have absolutely the same composition as Laugier and Nordenskiöld found for paranthine.[1] Laugier, Gehlen and Thomson obtained other results, all agreeing very nearly with the composition of 1 atom of bisilicate of lime + 2 atoms of silicate of alumina.

Alone, in the matrass, they give off a little moisture, without losing their transparency, which they retain till they begin to intumesce and fuse by exposure to a high temperature. Koupholite exhales an empyreumatic odour and blackens slightly. This is occasioned by most specimens from mineralogical collections having their interstices, by long exposure to the air, filled with dust,

---

[1] The text in the French translation, generally very accurate, is, in this instance, obscure; it is as follows:—" D'après l'analyse de Klaproth, ces fossiles auraient la composition chimique et jusqu' à la petite quantité d'eau que Laugier et Nordenskiöld out trouvée dans le paranthine." I have given the passage, through the kind assistance of a friend, from the German. C.

which chars by the heat. When this adventitious dust is entirely consumed, the crystalline laminæ of the mineral recover their transparence. Both the minerals give a blebby white glass.

With borax fuse readily into a diaphanous glass, which becomes turbid and very difficult to melt when saturated.

With salt of phosphorus decompose, and leave a flawy silica skeleton. The glass is transparent, and becomes opaline on cooling.

With soda fuse very laboriously, with incessant intumescence, into a scarce liquid, transparent glass; this result requires a prolonged blast. With a large quantity of soda the glass is at last converted into a semi-vitreous scoria.

17. *Paranthine, Scapolite.*[1]

(*a*). *Scapolite,* from Pargas.

Alone, in the matrass, gives off a little moisture, but does not lose its transparence.

On charcoal, not altered by moderate ignition; in a strong heat, fuses, intumesces violently, and forms, after the motion has ceased, a colourless, flawy, irregular and semi-translucid mass, which is no longer fusible.

With borax fuses, with continued effervescence, into a transparent glass. The effervescence ensues, even if the assay have been fused separately, till all internal motion has ceased, before it is treated with borax.

[1] One atom of silicate of lime + 3 atoms of silicate of alumina. C.

Salt of phosphorus decomposes it in like manner with effervescence.

With soda fuses slowly into a transparent glass, which a larger quantity of the flux renders difficult of fusion.

With solution of cobalt gives a blue glass.

(*b*). *Scapolite*, from Malsjo.

Alone first turns milk white, then fuses into a colourless glass, rendered turbid by the blebs it contains.

With borax, and salt of phosphorus, dissolves with effervescence like the preceding mineral.

With soda fuses very readily into a diaphanous glass, which a larger quantity of soda neither renders intumescent nor infusible.

(*c*). *Scapolite*, from Arendal (vitreous paranthine of Haüy).

Behaves precisely like the preceding mineral.

(*d*). *Dipyre*, from Mauléon, behaves just like the scapolite from Malsjö. Like the rest it gives off a small quantity of water without its transparence being affected, and appears to be a true scapolite, notwithstanding Vauquelin's analysis.

18. *Scolezite*.[1]

Alone, in the matrass, gives off water, and turns milk white; intumesces and fuses with difficulty into a blebby, colourless glass.

With borax, salt of phosphorus, and soda, behaves like the preceding, particularly the two last.

---

[1] One atom of trisilicate of lime + 3 atoms of silicate of alumina + 1 of water. C.

Remark.—M. Nordenskiöld found, amongst the minerals from Pargas, an anhydrous Scolezite. (See his Bidrag, &c., I. 67.)

19. *Mealy zeolite.*[1]
20. *Chabasie.*[2]
21. *Stilbite.*[3]
22. *Laumonite.*[4]

All these minerals behave like scolezite and prehnite; the blowpipe is incapable of detecting in them those differences which belong only to slight variations in the proportions of elements of the same nature. Laumonite gives in the first instant of fusion, a white globule, similar to enamel; but by a stronger heat, the globule itself is converted into a semi-transparent glass, with a turbid aspect from its blebby texture. It appears to be a general character of the double silicates of lime and alumina, particularly when they contain two, three, or a greater number of atoms of alumina to one of lime, to intumesce when they begin to fuse, as we have seen in all these silicates from epidote, inclusive, and as we shall see again in the calcareous tourmalines.

---

[1] One atom of trisilicate of lime + 3 atoms of bisilicate of alumina + 3 atoms of water. C.

[2] The same as mealy zeolite, but with a double quantity (6 atoms) of water. C.

[3] One atom of trisilicate of lime + 3 atoms of trisilicate of alumina + 6 atoms of water. C.

[4] One atom of bisilicate of lime + 4 atoms of bisilicate of alumina + 6 atoms of water. C.

23. *Cymophane, Chrysoberyl*, from Ceylon and Connecticut.[1]

Alone, no change. If previously pulverised, the edge of the cake assumes a vitreous appearance in a strong heat, but does not fuse.

With borax fuses slowly into a transparent glass, which remains diaphanous at every degree of saturation.

With salt of phosphorus fuses slowly, without residuum, into a transparent glass, which does not become opaline on cooling.

Soda has little action on cymophane; it causes no intumescence, the surface of the assay merely becomes dull. This flux has no greater action on the pulverised mineral.

With solution of cobalt the cake formed by kneading the pulverised mineral developes a fine blue colour without fusing.

24. *Lievrite, Yénite*, from the isle of Elba.[2]

Alone, in the matrass, gives off water free from acidity, which appears to be merely a mechanical

---

[1] One atom of sub-quadri-silicate of lime + 18 atoms of sub-quadri-silicate of alumina. C.—This composition, calculated on Klaproth's analysis, Beytrag, I. 102, does not appear to me to be correct.* I think the lime is not an essential element of the mineral, whose characters accord so perfectly with those of a subsilicate of alumina, that I doubt not we should place it beside dysthène in the family of aluminous minerals. B.

[2] One atom of silicate of lime + 4 atoms of silicate of protoxide of iron. C.

---

* The reader will probably be of the same opinion. C.

mixture, since the aspect of the mineral is not affected by its expulsion.

On charcoal fuses readily into a black globule, which becomes vitreous in the exterior flame; in the interior flame its surface becomes dull, and it acquires the property of being attracted by the magnet, provided the globule has not been heated to redness.

With borax dissolves readily into a dark, almost opaque glass, coloured by iron.

With salt of phosphorus decomposes, leaves a silica skeleton, and gives a glass strongly coloured by iron.

With soda lievrite fuses into a black glass. On platina foil it exhibits traces of manganese.

25. *Aplome*, calcareous garnet.

This is another instance of the same crystalline form extending to combinations so different, that their identity of figure cannot justify our considering them as substances belonging to the same species; we shall, however, in this, as in the case of amphibole, pyroxene and epidote, see all our difficulties satisfactorily removed by Mitscherlich's discovery of the *isomorphism* (l'isomorphisme) of certain bases, and the property of isomorphous combinations of crystallizing simultaneously, without being confined to fixed proportions. I have already said, that the protoxides of iron and manganese form one self-same class of isomorphous bases with lime and magnesia. Mitscherlich has shown that alumina, and the peroxides of iron and manganese, also compose a class of isomorphous bodies, but whose form is not the same as that of the bases of the first class; now it is these seven bases, just named, which occur in the garnets. We have seen that aluminiferous garnet is composed of 1 atom of silicate of protoxide of iron +1 atom of silicate of alumina. But it is clear that each of the three bases, isomorphous with protoxide of iron, may be substituted for it, without changing the form of the crystal. For instance, 1 atom of silicate of lime+1 atom of silicate of alumina will form a garnet just as well as the preceding compound. On the other hand, peroxide of iron and alumina being isomorphous, the first of these bases may take the place of the second; we can, therefore, again imagine garnets composed of 1 atom of silicate of lime+1 atom of silicate of peroxide of iron. These different garnets may occur in nature either separately, or crystallized in common, and in all sorts of proportions; and as their localities vary, the proportions of their constituent parts will vary also, and perhaps in toto, so that two different places will scarcely ever present the same species. This is completely confirmed by experiment. Bucholz analysed a garnet from Thuringia, which I have not had an opportunity of seeing; his analysis gives 1 atom of silicate of lime+1 atom of silicate of peroxide of iron. Melanite, analysed by Klaproth, is formed pretty nearly of 1 atom of silicate of lime+1 atom of silicate of alumina, with three atoms of silicate of lime, and silicate of peroxide of iron (equivalent altogether to 1 atom of silicate of alumina+3 atoms of silicate of peroxide of iron+4 atoms of silicate of lime), and there is no reason for supposing that the combination which takes place in this instance between the two double silicates, is more intimate or more *chemical* than those which Mitscherlich formed at pleasure in his beautiful experiments on the simultaneous crystallization of isomorphous sulphates, with variable proportions of the different sulphates. According to Murray's analysis of Dannemora garnet, it is composed of three others, which are, respectively, 1 atom of silicate of lime+1 atom of silicate of alumina...1 atom of silicate of protoxide of manganese+1 atom of silicate of alumina...and 1 atom of silicate of protoxide of iron+1 atom of silicate of alumina. The analysis of a deep brown garnet, from Longbanshytta, (*Rothoffite*,) by Rothoff, teaches us that it is formed of two others, namely, 1 atom of silicate of lime+1 atom of silicate of peroxide of manganese...and 1 atom of silicate of lime+1 atom of silicate of peroxide of iron, the second being nearly in three times greater quantity than the first. What I have said is sufficient to show how it is that garnets differ so much from one another in their chemical composition. As this composition almost always varies with the places they come from, it seems to me more convenient to distinguish them by their localities, than to invent peculiar names for each variety, which in time would form a very long list, uselessly burthensome to the memory.

(α). *Black garnet*, from Frescati. (Melanite.)

Alone fuses without foam into a black brilliant globule.

With borax fuses slowly and difficultly into a glass coloured by iron.

With salt of phosphorus decomposes slowly, and leaves a silica skeleton. The tint from iron disappears on cooling.

With soda fuses into a black glass globule; an increased dose of the flux renders it more difficult to fuse; nevertheless, it still does fuse after the excess of soda has been absorbed by the charcoal. On platina foil it shows traces of manganese.

(β). *Green garnet*, from Sala.

Fuses, with strong intumescence, into a black brilliant glass.

With borax and *salt of phosphorus* behaves like the preceding.

With soda decomposes and intumesces, but afterwards fuses into a black brilliant globule. On platina foil exhibits traces of manganese.

(γ). *Clear brown garnet*, from Dannemora.

Fuses readily without intumescence into a black brilliant globule.

With borax fuses into a transparent glass, which in the oxidating flame assumes a false amethyst colour, and, in the reducing flame, the green colour, characteristic of iron.

With salt of phosphorus behaves like the preceding.

With soda, at first becomes green, then intumesces and fuses into a black globule with metallic lustre.

(*δ*). *Deep brown or black garnet,* from Longbanshytta. (Rothoffite.)

Alone fuses with difficulty into a black, dull, or semi-vitreous globule.

With borax fuses with difficulty into a dark green glass.

With salt of phosphorus behaves like the preceding minerals.

With soda fuses with difficulty into a black glass. On platina foil Rothoffite shows that it is much loaded with manganese.

(*ε*). *Red garnet,* from the limestone quarry at Kulla in Finland; Romanzowite.

Alone becomes brown by heat without losing its transparence; fuses with slight intumescence into a greenish, blebby glass globule.

With borax dissolves with extreme difficulty. With a prolonged blast it first becomes green on the edges, then in the central part, which now shines with a more intense green colour than the surrounding glass; at length it totally fuses.

With salt of phosphorus as with the preceding minerals.

With soda intumesces and fuses into a green glass, which more soda renders difficult to fuse. On the platina foil gives traces of manganese.

Remark.—This garnet much resembles the essonite from Ceylon, both in aspect and composition. According to Nordenskiöld, romanzowite is composed of 1 atom of silicate of peroxide of iron+3 atoms of silicate of lime+5 atoms of silicate of alumina; and the analysis of Klaproth gives the composition of essonite from Ceylon as 1 atom of silicate of peroxide of iron+4 atoms of silicate of lime+5 atoms of silicate of alumina. Both occur in limestone.

(ζ). *Allochröite,* from Berggieshübel, in Saxony.

Alone fuses easily without intumescence into a black brilliant glass.

With borax fuses easily into a glass coloured green by iron.

With salt of phosphorus and soda behaves like the preceding minerals. On platina foil gives slight traces of manganese.

26. *Essonite.*

(*a*). *From Ceylon,* Kaneelstein.

Alone does not turn brown by heat; fuses readily into a glass, which at first has the same colour as the mineral, but afterwards becomes greenish; in both cases it is transparent.

With borax fuses readily into a transparent glass, very slightly coloured by iron.

With salt of phosphorus behaves like the preceding.

With soda fuses into a green glass, which a larger dose of the flux converts into a difficultly fusible grey scoria.

(*b*). *From Brasil.*

Its dark red colour does not become darker by heat; fuses readily into a black globule with a metallic surface, but not very attractable by the magnet.

With borax fuses with the greatest facility into a glass coloured by iron.

With salt of phosphorus decomposes like the other garnets.

With soda decomposes; a very small quantity of soda gives a black glass, a larger quantity converts it into an infusible scoria.

27. *Idocrase.* Haüy has separated this mineral and the preceding, which differ from the garnets in their primitive form, into two distinct species. They are, however, but mixtures of the same silicates in different proportions, and what we have said respecting garnets applies to them. Time will show if any other essential difference besides form exist between them.

(*a*). *Vesuvian*, from Vesuvius and Fassa.

Fuses very easily, with intumescence, into a dark glass, which, in the exterior flame, becomes yellow and transparent.

With borax fuses easily into a diaphanous glass coloured by iron.

With salt of phosphorus decomposes easily, and leaves a silica skeleton. The glass becomes opaline on cooling.

With soda vitrifies with greater difficulty than the garnets; a fresh quantity of soda converts the glass into scoria.

Remark.—The facility with which the idocrases dissolve in borax and glass of phosphorus, is connected with their property of intumescing at a certain temperature, for when this takes place in presence of the fluxes, it affords the latter a greater number of channels by which to penetrate into the interior of the assay, and thus enables them to act at once on an increased number of points. As to the difficulty with which the idocrases dissolve with soda, it must be attributed to their containing much less iron than the garnets.

(*b*). *Egerane,* from Egra.

Fuses with intumescence, but without first becoming opaque, into a greenish, blebby glass.

With the fluxes behaves like the preceding.

(*c*). *Loböite,* magnesian idocrase, from Gökum and Frugord.

Alone becomes opaque, splits, and then fuses readily, with intumescence, into a green or yellowish globule, which seems to be composed of heterogeneous parts.

With borax fuses almost instantly with slight intumescence.

With soda and salt of phosphorus behaves like the preceding minerals.

(*d*). *Cyprine,* cupreous idocrase, from Tellemarken, in Norway.

A moderate ignition does not alter its fine blue colour. Whilst hot it appears black, but it becomes blue again on cooling. It fuses easily, with violent intumescence, into a blebby globule, which is black in the oxidating flame, but in the reducing exhibits a red colour, derived from protoxide of copper.

With borax fuses easily into a diaphanous glass, which in the oxidating flame becomes green. In the reducing flame it becomes colourless, and, if not sufficiently saturated with cyprine, the effect of the copper cannot be developed without the addition of tin.

With salt of phosphorus decomposes immediately, the assay intumesces, and forms a flawy mass, which is green when cold, and becomes red on the surface, in the interior flame. A large quantity of the salt developes the green colour of copper, but the red can only be obtained by the help of tin.

With soda a black glass, which takes a larger dose of soda than the preceding. By reduction, the assay gives a good deal of copper.

Remark.—Mitscherlich has shown that oxide of copper belongs to the same class of isomorphous bases as lime, magnesia, protoxide of iron, &c.; but instances of the substitution of the former for either of the others are extremely rare in the mineral kingdom. Cyprine, however, is one of them.

28. *Pyrope.*

(*a*). *Pyrope,* from Ceylon.

Becomes brown by heat, and, at last, black and opaque; if we now suffer the assay to cool, and watch its appearance whilst cooling by daylight, we may perceive it become successively dark green, fine chrome green, and colourless; and, lastly, the fine fiery red that distinguishes it in its natural state, re-appears in full vivacity. It fuses with difficulty and without intumescence, into a brilliant black glass.

With borax fuses into a chrome green glass, the beauty of whose colour depends on the degree of saturation.

With salt of phosphorus decomposes as slowly as possible, and leaves a silica skeleton. Before the assay decomposes, the glass assumes a green colour. The assay retains its original state and red colour for a long time, till at last it is gradually converted into the silica skeleton. The globule becomes opaline on cooling, and assumes a chrome green colour.

With soda the assay decomposes, but does not dissolve, at least only in very minute quantity. The result of the decomposition is a red brown globule composed of scoriæ.

(*b*). *Bohemian Pyrope.*

Becomes black and opaque by heat; seen by refracted light, during its cooling, it first appears dirty yellow; immediately afterwards it becomes red; lastly, it recovers its original colour.

With borax fuses into a glass strongly coloured by iron, without any notable mixture of chrome green.

With salt of phosphorus behaves like the preceding, but gives a lighter green colour.

With soda behaves like the preceding.

29. *Axinite*, Thumerstein, from Dauphiny.

Alone intumesces and fuses readily into a dark green glass, which blackens in the exterior flame. The black colour is developed by the peroxide of manganese.

With borax fuses readily into a glass coloured by iron, but which in the exterior flame assumes an impure amethystine tint.

With salt of phosphorus decomposes with the common phenomena.

With soda first becomes green, then fuses into a black glass with an almost metallic lustre.

30. *Anthophyllite,* from Greenland.

Alone unalterable and infusible both in fragment and in powder.

With borax fuses with difficulty into a glass coloured by iron.

With salt of phosphorus decomposes slowly, and leaves a silica skeleton.

With soda fuses with difficulty into a scoriaceous mass; gives no trace of manganese.

31. *Gehlenite,* from Monzoni, in the Valley of Fassa.[1]

Alone, no change.

With borax dissolves with extreme difficulty into a glass faintly coloured by iron.

With salt of phosphorus becomes gradually

[1] Two atoms of silicate of lime + 1 atom of subsilicate of alumina; about ¼ of the latter is replaced by subsilicate of peroxide of iron. The formula calculated on the authority of Fuchs. C.

transparent on the edges, and dissolves entirely without previous intumescence.

With soda intumesces, but does not fuse.

With solution of cobalt gives a dark, impure blue.

With boracic acid fuses into a transparent glass, coloured green by iron; gives no phosphuret by the addition of metallic iron.

32. *Cerine,* from Bastnäs.[1]

Alone, in the matrass, gives off a little water without altering its appearance; the water cannot, therefore, constitute one of the chemical elements of the mineral; fuses easily with intumescence into a black, shining glass globule.

With borax fuses easily into an opaque black glass, which becomes blood red in the exterior flame, and retains that colour whilst hot, but changes it for a more or less dark yellow when cold. In the reducing flame, the glass exhibits a fine iron green colour. It does not become opaque by *flaming*.

With salt of phosphorus decomposes, and leaves an opaque silica skeleton. Whilst hot, the glass exhibits the colour indicative of iron, but becomes colourless and opaline on cooling.

With soda fuses into a black glass, which is not rendered less fusible by a larger quantity of the flux.

---

[1] The formula gives 1 atom of silicate of lime + 2 atoms of silicate of alumina, mixed with a considerable quantity of silicate of protoxide of cerium, and silicate of protoxide of iron. C.

Remark.—I have not had an opportunity of assaying the *allanite* of Thomson by the blowpipe. I conceive it must behave very much like cerine.[1]

[1] By the kindness of Mr. Heuland, who neglects no opportunity of assisting experimental research, I am enabled to supply the deficiency. That gentleman had the goodness to furnish me with two small imperfect crystals of allanite, from Alluk, near Kakasoeits, in South Greenland, which he received from Sir Charles Giesecké, and which are unquestionably of the same species as that analysed by Dr. Thomson. The following are the results of my experiments.

Alone, in the matrass, decrepitated slightly, and gave off moisture, which condensed into drops of pure water. The colour of the assay became lighter, and somewhat greenish yellow on the surface.

In the platina forceps a small fragment, with a very thin edge, merely became greenish yellow on the surface, but did not fuse with a very intense heat continued for a considerable time. *On charcoal,* same effect.

When pulverised, it was equally infusible.

Remark.—In this respect the specimen I examined differs from the allanite which Dr. Thomson analysed, who states it to have frothed before the blowpipe, and melted imperfectly into a black scoria.

With borax, on the platina wire, a small fragment of the assay was very little acted on; it revolved rapidly in the globule, slowly diminished in bulk, and became white; the glass was perfectly transparent, and presented a very light fine yellow colour. In the reducing flame, the remainder of the globule dissolved, glass still transparent, and green on cooling. When in powder, the assay dissolved on platina wire pretty readily; in the exterior flame, the glass, quite saturated, had a rather dark yellow colour whilst hot, which became lighter on cooling, was transparent, and could not be made opaque by flaming. In the reducing flame, the globule

33. *Orthite,* from Finbo and Gottliebsgong.[1]

Alone, in the matrass, gives off water, and at a high temperature its colour becomes clearer.

became opaque from excess of the assay, and dirty yellowish brown. With a further quantity of the flux, in the exterior flame, the original yellow colour re-appeared, which, in the reducing flame, assumed a somewhat greener tinge, but experienced no other change.

With salt of phosphorus, little action on a fragment of the assay; glass at first yellow; when cold, nearly colourless, with a very slight tint of green, and containing a rather compact silica skeleton, of a light reddish colour. The pulverised assay fused readily into a light bluish green glass, almost colourless on cooling. With the assay in excess, the glass became opaque and dull reddish purple. A fresh quantity of the flux restored the original colour. In the reducing flame, the colour became light dirty brown whilst hot, but on cooling re-assumed the bluish green.

With soda, on platina wire, or charcoal, no sensible action on a fragment of the assay, and scarcely more when pulverised; globule opaque and almost white, very slightly inclining to bluish green, which, in the reducing flame, disappeared; the globule became grey-white, scarcely inclining to reddish, bubbled considerably, projected little ignited particles, and gave off elastic matter, which burnt with continual flickering and enlargement of the flame of the lamp, whilst the mass rapidly diminished in volume till hardly any thing was left on the wire.

With soda, on platina foil, no trace of manganese.

Some other specimens which were given me for allanite, presented precisely the same phenomena as Berzelius obtained with cerine, except that with soda they gave, in the exterior flame, a dirty brown opaque mass containing black particles of the assay dispersed through it, instead of the black glass. They also gave less water when heated in the matrass than pure allanite, and their appearance was not altered by the

Alone, on charcoal, intumesces, becomes yellow brown, and at length fuses, with a brisk bubbling, into a black, blebby glass.

With borax dissolves readily. The glass, whilst hot, is blood-red; when cold, it is yellow. In the reducing flame, it assumes an iron tint.

With salt of phosphorus decomposes readily with the usual phenomena.

With soda intumesces; with a very small dose of soda the assay fuses; with a larger quantity it swells up, and is converted into a greyish yellow scoria. On the platina foil, traces of manganese are visible.

34. *Pyrorthite,* from Korarf.[2]

Alone, in the matrass, first gives off a very large quantity of water, the last portions of which are yellowish, and have an empyreumatic odour. The residual matter is as black as charcoal.

---

operation. All these specimens were, I believe, from Sweden, except one which was from Iglorsoit, in South Greenland, a new locality of allanite discovered by Giesecké, from whom Mr. Heuland received it. It is in a quartz matrix, on gneiss, and appears to be crystallized in rhombic prisms. C.

[1] One atom of silicate of lime + 3 atoms of silicate of alumina + 2 atoms of water, mixed with the silicates of the protoxides of cerium and iron. C.

[2] One atom of silicate of lime + 3 atoms of silicate of alumina + an unknown number of atoms of water; it contains nearly $\frac{1}{4}$ of its weight of carbon, and $\frac{1}{4}$ of its weight of water, besides a notable quantity of silicate of protoxide of cerium, with smaller quantities of the silicates of the protoxides of iron and manganese, and silicate of yttria. C.

Heated moderately *on charcoal*, and then to redness at one point, it takes fire, and continues to shine by itself without flame or smoke. If we collect several small fragments into a heap, or, having reduced the mineral to a coarse powder, form it into a little conical loaf, the combustion will ensue in a still more lively manner. A gentle blast makes the phenomenon more distinct. After roasting, the mineral is white, or grey-white; its tint then varies with the nature of the little pieces operated on, and sometimes inclines to red. These little pieces are so porous and light that we cannot keep them on the charcoal during the blast. Between the forceps, they fuse with difficulty into a black glass with an unpolished surface.

With borax pyrorthite fuses readily into a glass, which presents the same phenomena as are shown by the preceding mineral with the same flux.

With salt of phosphorus dissolves with difficulty. The porous piece remains on the surface of the globule, whilst it is in fusion, and sinks into it as it cools; if we heat it afresh, the porous assay reappears on the surface.

With soda the phenomena are the same as those with orthite.

7. *Strontium.*

1. *Sulphate of Strontita.*[1]

[1] One atom of strontita $52 + 1$ atom of sulphuric acid $40 = 92$. C.

The crystallized mineral decrepitates; sulphate of strontita fuses on charcoal, in the exterior flame, into a milky white globule, which, before the interior flame, spreads over the charcoal, decomposes, becomes infusible, and leaves a hepatic mass; after it is cold, this mass, when held near the nose, has a slight odour of rotten eggs; its flavour is hepatic and caustic: on platina foil it dissolves in great part in muriatic acid; if the solution be evaporated to dryness, and the salt be laid on a narrow slip of paper, moistened with alcohol, and set on fire, the flame in contact with the salt will be coloured red. This phenomenon ensues even with sulphate of baryta if it contain strontita.

With borax fuses with effervescence into a transparent glass, which becomes yellow or brown on cooling, and opaque if the proportion of sulphate be considerable.

With salt of phosphorus behaves like strontita.

With soda swells up, decomposes, penetrates the charcoal, and forms a strongly hepatic mass.

With soda and silica gives a glass coloured by the hepar.

With fluor spar fuses into a transparent glass, which becomes enamel white on cooling.

2. *Carbonate of Strontita.* (See Strontita, p. 78)

8. *Barium.*

1. *Sulphate of Baryta.*[1]

---

[1] One atom of baryta $78 + 1$ atom of sulphuric acid $40 = 118$. C.

2. *Triphane, Spodumène,* from Utö, and the Tyrol.[1]

Alone, in the matrass, gives off water, and becomes more turbid and whiter than it was at first.

On charcoal intumesces like the double silicates of lime and alumina, and afterwards fuses into a colourless, almost transparent glass.

With borax intumesces, but does not fuse easily; the tumified mass becomes transparent and globular, but for a long while remains undissolved.

With salt of phosphorus intumesces in a similar manner, decomposes pretty readily, and leaves a silica skeleton.

With soda swells up and fuses into a transparent glass, which, although rendered opaque by a larger quantity of soda, does not thereby become difficult of fusion.

With solution of cobalt gives a blue glass.

3. *Petalite,* from Utö.[2]

Behaves in all respects like feldspar (which see).

4. *Tourmaline,* from Utö.[3]

(*a*). *Red, and clear green.*

Alone turns milk white, intumesces a little, splits

---

[1] By the formula, calculated on A'rfwedson's analysis, it is composed of 1 atom of trisilicate of lithia + 3 atoms of bisilicate of alumina. C.

[2] Also from Arfwedson's analysis; 1 atom of sex-silicate of lithia + 3 atoms of trisilicate of alumina. C.

[3] One atom of silicate of lithia + 9 atoms of silicate of alumina? On the same authority. C.

obliquely, does not fuse, but becomes scoriaceous on the surface.

With borax first effervesces slightly, turns milky white, and then fuses, slowly and with difficulty, into a transparent, colourless glass.

With salt of phosphorus the effervescence, colouring and solution take place in the same manner, and without the assay dividing; at the same time its bulk diminishes. The glass produced becomes opaline on cooling.

With soda fuses with extreme difficulty into an opaque glass. On platina foil it assumes a dark green colour.

(*b*). *Clear blue, finely striated.*

Alone intumesces a little, whitens, does not fuse, but becomes scoriaceous on the surface, and blebby in the part most strongly heated.

With borax fuses pretty easily, with effervescence, especially if too much of the assay be not added at once, into a diaphanous glass.

With salt of phosphorus intumesces and effervesces; the skeleton divides, and afterwards in great measure dissolves. The globule becomes opaline on cooling.

With soda fuses with difficulty into a dark glass, whose fusibility is diminished, but not destroyed, by a further quantity of soda. On the platina foil traces of manganese are perceptible.

(*c*). *Dark blue, in large crystals.* Indigolite.

Alone swells up very much, particularly longitudinally, so that in this direction it increases to nearly three times its original dimensions; but the increase of volume only taking place on one side, the assay becomes curved, and rolls over; at the same time it is converted into a black scoria.

With the fluxes behaves like the preceding mineral.

Remark.—The varieties *b* and *c* appear to result from a mixture of the lithium-tourmaline with that species of tourmaline, which I have placed further on in the potassium family.

5. *Lepidolite,* from Roscena, and Utö. (Lithionglimmer.)?[1]

Alone, in the matrass, gives off water, which, if the heat be pushed to redness, is sensibly loaded with fluoric acid, yellows brazil wood paper, and dulls the glass here and there by the silica it deposits on its surface.

On charcoal intumesces and fuses very readily into a transparent, colourless, blebby glass globule.

With borax fuses readily, and in large quantity, into a transparent glass.

With salt of phosphorus decomposes and leaves a silica skeleton; the globule becomes opaline on cooling.

With soda fuses readily, with intumescence, into a slightly blebby transparent glass.

With solution of cobalt becomes blue in fusing.

---

[1] Professor C. Gmelin, of Thübingen, found lithia and potassa in this mineral. If the difference between it and common mica be owing to its containing lithia, the presence of potassa may be derived from a mixture of common mica. B.

With boracic acid and iron gives no phosphuret of iron.

10. *Sodium.*

1. *Sulphate of soda.*[1]

Alone, in the matrass, fuses in its water of crystallization, which evaporates. The dried salt fuses on charcoal, penetrates it, and is converted into sulphuret.

Fused with soda it passes into the charcoal, by which it is distinguished from salts with earthy bases. With soda and silica, it gives a glass coloured by the hepar.

2. *Glauberite,* from Villarubia, in Spain. I am indebted to M. Brongniart for the specimen used in these experiments.[2]

Alone, in the matrass, decrepitates violently, and gives off a very little water. In an incipient red heat it afterwards fuses into a transparent glass, which gives off no volatile substance.

On charcoal whitens on the first impulse of the heat, and then fuses easily into a clear globule, which loses its transparence on cooling. In the reducing flame, it fixes and becomes hepatic. The sulphuret of soda penetrates into the charcoal, and the lime remains on the surface in the form of a very porous white globule.

---

[1] One atom of sulphuric acid 40 + 1 atom of soda 32 + 10 atoms of water 90 = 162. C.

[2] One atom of sulphate of soda 72 + 1 atom of carbonate of lime 50 = 122. C.

With borax dissolves with brisk effervescence; the mass is absorbed by the charcoal.

With salt of phosphorus fuses with effervescence into a milk white glass.

With fluor spar fuses like gypsum.

With soda is decomposed; a hepatic mass passes into the charcoal, and the lime remains on the surface. With soda and silica forms a glass coloured by the hepar.

3. *Sea salt.*[1]

Alone, in the matrass, decrepitates, and gives off a little water.

On charcoal fuses and is absorbed, with the disengagement of fumes. *On the platina foil* fuses into a diaphanous mass, which loses its transparence on congealing. *With salt of phosphorus, impregnated with oxide of copper*, we obtain the fine blue flame characteristic of muriatic acid. *With soda*, on platina foil, it dissolves without becoming turbid.

4. *Borax* or *Tincal*.

Intumesces like borax, carbonizes, gives off an empyreumatic odour, and then fuses into a transparent globule.

5. *Cryolite*, from Greenland.[2]

---

[1] One atom of sodium 24 + 1 atom of chlorine 36 = 60. C.

[2] The analyses of Klaproth and Vauquelin very nearly agree, except in the quantity of fluoric acid and water contained in cryolite, of which the former found 40, the latter 47 per cent. By Klaproth's analysis, the alumina is to the soda as 24 : 36. The formula, calculated on what data does not appear, gives

Alone, in the matrass, gives off a little water, and decrepitates, but does not lose its transparence.

In the open tube, the flame being directed into it and immediately on the assay, the glass is strongly attacked, and the moisture which condenses in the tube, indicates, by its effects, the presence of fluoric acid.

On charcoal fuses into a transparent globule, which becomes opaque on cooling. By a continued blast, the glass flows abroad, the fluate of soda is absorbed by the charcoal, and an aluminous crust remains on the surface.

Borax readily dissolves a large quantity of cryolite, and converts it into a transparent glass, which becomes milk white on cooling. *Salt of phosphorus* produces the same effects. The glass globule sometimes assumes a reddish tint from the presence of a small quantity of copper.

With soda fuses into a clear glass, which flows abroad, and becomes milky white on cooling.

With boracic acid and iron we obtain no phosphuret of iron.

6. *Sodalite.*

(*a*). *Sodalite,* from Vesuvius.[1]

(I had the specimen from M. Haüy.)

---

the composition of cryolite as 3 atoms of fluate of soda + 1 atom of sesqui-fluate of alumina. C.

[1] One compound atom, composed of an atom of chloride of sodium, with 2 atoms of submuriate of alumina + 4 compound atoms, each composed of an atom of silicate of soda, with three atoms of silicate of alumina. Calculated from Arfwedson's analysis. C.

Alone, in the matrass, gives no traces of water.

On charcoal, no change; but, by a very powerful blast, its edges become rounded, without intumescence, or the formation of bubbles, and without the assay losing its transparence.

With borax fuses in small quantity, and with extreme difficulty, into a transparent, colourless glass.

With salt of phosphorus does not intumesce; fuses with difficulty in small quantity without being decomposed. The glass becomes opaline on cooling.

With a small quantity of soda gives a transparent glass, surrounding an untouched nucleus. A larger quantity of soda decomposes the mineral, causes it to swell up, and become infusible. If we then add a further dose of the flux, the intumesced mass fuses into a turbid or opaque, but colourless, glass.

With solution of cobalt the fused edges are coloured blue.

(*b*). *Sodalite,* from Greenland.[1]

(The specimen was given me by M. Cordier.)

Alone, in the matrass, gives off a little water, but the appearance and transparence of the mineral remain unaltered.

On charcoal fuses with very brisk intumescence and ebullition, into a distorted, colourless glass.

---

[1] The formula, calculated from Thomson's analysis, gives 1 atom of silicate of soda + 2 atoms of silicate of alumina. By the ? annexed, the author seems doubtful of its accuracy. C.

With borax behaves like the preceding.

With salt of phosphorus decomposes with the greatest difficulty. After having continued the blast for some time, the edges of the assay become siliceous. The glass turns opaline on cooling.

With soda vitrifies much more difficultly than the vesuvian sodalite. The glass is opaque.

With solution of cobalt behaves like the preceding.

7. *Lapis-Lazuli.* The specimen used in the following experiments was very pure, and had a natural cleavage. I am indebted for it to M. Cordier.

Alone, in the matrass, gives off a little water, without changing its aspect, or losing its transparence.

On charcoal fuses with difficulty into a white glass; in the first moments of fusion the white is mixed with blue, but, on continuing it, the colour entirely disappears. The unfused portion preserves here and there some blue spots, and becomes dark green in the parts near the fused portion. The lapis lazuli, whose texture is not lamellar, fuses more easily, with slight intumescence.'

With borax fuses with continued effervescence into a transparent, colourless glass. During the solution, the undissolved nucleus shines with greater brightness than the glass which surrounds it.

' Is not this the common lapis lazuli, the lazurstein or azurestone of Werner, and the former the lazulite or azurite of the same mineralogist? C.

With salt of phosphorus the fusion is also accompanied by a prolonged effervescence, and the same phenomena of ignition; the solution is complete, and leaves no silica; the glass is colourless, and becomes opaque on cooling.

With soda fuses partially into an opaque greenish grey glass, which, on cooling, assumes a red colour, similar to that produced by a hepar. With a larger quantity of soda, the phenomena are the same.

Remark:—The effervescence with the fluxes, as well as the developement of the hepar colour by soda, seem to indicate the presence of sulphuric acid.

8. *Mesotype.*[1]

Alone, in the matrass, gives off water.

On charcoal the radiated species expands longitudinally, and the compact intumesces; both fuse afterwards into a blebby colourless glass. The mesotype, in large crystals, merely becomes opaque, and then vitrifies without intumescence.

With borax fuses with difficulty. The assay intumesces indeed with glass of borax, but a portion remains as a white mass, and requires a very long blast for its solution. The glass is transparent and colourless.

With salt of phosphorus decomposes very easily, and leaves a silica skeleton. The glass becomes opaline on cooling.

[1] One atom of trisilicate of soda + 3 atoms of silicate of alumina + 2 atoms of water. By the tables in the Nouveau Systême, the formula is calculated on Klaproth's analysis, Beytr. v. 49. C.

With soda fuses into a transparent glass.

Remark.—The yellowish compact mesotype, from Miss, in Bohemia, becomes red before it fuses. This phenomenon is owing to the mineral being coloured by a hydrate of peroxide of iron.

9. *Mesolite*, from Hauenstein, in Bohemia.[1]

Behaves like the preceding.

10. *Albite*, from Broddbo, Finbo, and Haddau, in Connecticut.[2]

Behaves in all respects like feldspar. (See the article relating to that mineral.)

11. *Analcime*, from Fassa, and Etna.[3]

Alone, in the matrass, gives off water. The transparent variety becomes milky white.

On charcoal, little change of aspect by a moderate heat; in a stronger heat it becomes transparent, and then fuses into a slightly blebby, diaphanous glass, without previous intumescence or bubbling.

With borax dissolves, even in powder, with great difficulty, and leaves an opaque, concrete residuum. The vitreous portion is transparent.

With salt of phosphorus decomposes slowly, and gives a slightly blebby skeleton. The glass is transparent, and does not become opaline till after

---

[1] The formula seems to be calculated on Gehlen and Fuch's analysis (Nouv. Syst.); it represents the mineral as composed of 1 atom of trisilicate of soda + 1 atom of trisilicate of lime + 6 atoms of silicate of alumina + 3 atoms of water. C.

[2] One atom of trisilicate of soda + 3 atoms of trisilicate of alumina. C.

[3] One atom of bisilicate of soda + 3 atoms of bisilicate of alumina + 3 atoms of water. From Rose's analysis. C.

a long continued blast, and even then the effect is scarcely perceptible.

With soda gives a transparent glass.

With solution of cobalt gives a blue glass.

12. *Ekebergite,* natrolite from Hesselkulla.[1]

Alone, in the matrass, gives off a little water, but its appearance does not change.

On charcoal whitens, loses its transparence, intumesces a little, and then fuses into a blebby, colourless glass.

With borax, and salt of phosphorus, dissolves with effervescence, precisely like paranthine or scapolite.

With soda fuses, like the paranthine from Pargas, with great difficulty, into a transparent glass, whose greenish colour is developed by iron.

13. *Rubellite,* tourmaline apyre (soda-tourmaline), from Siberia and America.[2]

Alone, in the matrass, gives off a little moisture; no change of aspect; the transparent variety gives off no water.

On charcoal turns milk white, intumesces much more than the tourmaline from Utö, splits like it

---

[1] From Ekeberg's analysis, 1 atom of bisilicate of soda + 3 atoms of bisilicate of lime + 12 atoms of silicate of alumina. C.

[2] *Red tourmaline.* The magnificent specimen of this substance in the British Museum, was presented to Col. Symes by the King of Ava, and afterwards deposited in Mr. Greville's cabinet; it was valued at 500*l.* by the Commissioners appointed by Parliament to value that gentleman's collection, previously to its being purchased by Government. By the formula it is 1 atom of silicate of soda + 9 atoms of silicate of alumina? C.

in an oblique transverse direction, does not fuse, but vitrifies on the edges.

With borax fuses easily, and with effervescence, into a transparent glass, in which some flocculi, that dissolve slowly, may be seen floating.

With salt of phosphorus decomposes pretty easily, with effervescence, into an opaline glass, and leaves a silica skeleton.

With soda fuses, with great difficulty, into an opaque glass.

On platina foil exhibits the effects of manganese in an intense degree; the transparent, colourless tourmaline, from America, is more particularly remarkable in this respect.

Remark.—We see that the tourmaline apyre is more soluble with the fluxes than the tourmaline from Utö, apparently because the first intumesces at a high temperature more than the second, which, as we have observed before, favours the penetration, and, consequently, the chemical action of the solvents.

11. *Potassium.*

1. *Polyhalite,* from Ischel, in Austria.[1]

Alone, in. the matrass, gives off water, and its red colour fades.

On charcoal fuses into an opaque, reddish yellow globule, which, in the interior flame, congeals,

[1] One atom of sulphate of potassa + 2 atoms of sulphate of lime + 1 atom of sulphate of magnesia + 4 atoms of water; according to Professor Stromeyer. C.

whitens, and leaves an empty shell; its flavour is then saline and slightly hepatic.

With borax fuses, with brisk effervescence, after a somewhat long blast, into a diaphanous glass, which, on cooling, becomes dark red without losing its transparence. A large quantity of the assay is necessary for the glass to become opaque on cooling.

With salt of phosphorus fuses into a transparent, colourless glass. The assay must be in large proportion to render the glass opaque.

With soda decomposes, leaves an earthy mass, which, in the reducing flame, assumes a yellowish colour from a mixture of hepar.

With fluor spar fuses into an opaque globule.

2. *Alum.*[1]

In the matrass fuses, intumesces, and gives off water. The dry mass gives off sulphurous acid at a red heat, but no sublimate. The residuum behaves with the fluxes like alumina.

3. *Alaunstein* (alum-stone), from Tolfa. According to Cordier's analysis, it is a siliceous subsalt of potassa, alumina, and sulphuric acid, in which the oxygen of the alumina is to that of the potassa as 15 : 1.

Alone, in the matrass, first gives off water; in a stronger heat sulphate of ammonia sublimes.

On charcoal, in a strong heat, contracts without fusing.

With borax fuses, with effervescence, into a colourless, transparent glass.

---

[1] One atom of sulphate of potassa + 2 atoms of sulphate of alumina + 24 atoms of water. C.

With salt of phosphorus fuses pretty readily, and leaves a semi-transparent silica skeleton; the glass does not become opaline on cooling.

With soda does not fuse.

With solution of cobalt gives a fine blue.

4. *Saltpetre.*[1]

Alone, in the matrass, gives off a little moisture, and fuses below a red heat.

On charcoal detonates at the moment it fuses, and leaves an alcaline mass in the charcoal.

5. *Amphigène,* Leucite.[2]

Alone, in the matrass, gives off no water.

On charcoal, no change, nor fusion, even in powder. If the pulverised mineral be mixed with a very little carbonate of lime, the mixture fuses very evidently.

With borax fuses slowly, but in large quantity, into a diaphanous glass.

Salt of phosphorus has little action on amphigène, either in fragment or in powder; nevertheless, it converts it into a transparent globule, having nearly the same degree of refrangibility throughout, so that no undissolved portion can be perceived without minute inspection. We may satisfy ourselves of it, by compressing the liquid globule between two cold bodies.

With soda fuses slowly, with effervescence, into a transparent, although blebby glass.

[1] One atom of potassa 48 + 1 atom of nitric acid 54 = 102. C.
[2] One atom of bisilicate of potassa + 3 atoms of bisilicate of alumina. The formula calculated from Arfwedson's analysis. C.

With solution of cobalt the assay gives a fine blue, but does not fuse.

6. *Meionite*, the dioctohedral variety, from Vesuvius.[1]

Alone, in a thin spangle, it throws out, at certain points, a blebby foam; the whole mass soon afterwards bubbles up, and the bubbling lasts a long time. The result is a blebby, colourless glass.

With borax fuses slowly, with prolonged effervescence, into a transparent glass.

With salt of phosphorus decomposes with effervescence, and gives a siliceous residuum, and a glass which becomes opaline on cooling.

With soda fuses slowly, with much intumescence, into a transparent glass. A large dose of soda is requisite, and the assay, for a long while, retains an opaque side.

With solution of cobalt merely fuses on the edge, which is coloured blue.

Remark.—These experiments were made on the specimen analysed by Arfwedson. (Afh. i Fysik, &c. vi. 255.) I must here inform the reader that Professor Leop. Gmelin has analysed a substance called meionite, whose composition is altogether different. (Schweiger's Journal, xxv. p. 36.)

7. *Felspar.*[2]

Alone, in the matrass, transparent felspar gives

---

[1] One atom of trisilicate of potassa + 3 atoms of bisilicate of alumina. From Arfwedson's analysis. C.

[2] One atom of trisilicate of potassa + 3 atoms of trisilicate of alumina. From Arfwedson's analysis. C.

off no water. The cracked, opaque felspar often affords a large portion of water, which was contained mechanically in the interstices of the mineral.

On charcoal, in a bright heat, it becomes vitreous, semi-transparent and white, and fuses with difficulty on the edge into a blebby, semi-transparent glass. It is a mineral of very difficult fusion.

With borax fuses very slowly, without effervescence, into a diaphanous glass.

Salt of phosphorus attacks it with great difficulty; with the pulverised mineral, it gives a globule which becomes opaline on cooling, and leaves a silica skeleton.

With soda the solution is slow, and attended with effervescence; it gives a transparent glass, very difficult to fuse and obtain free from blebs.

With solution of cobalt only the fused edges are coloured blue.

Remark 1*st.*—The iridescent felspar, from Labrador, in America, according to Klaproth's analysis, (Beytr. vi. 255,) should be composed of 1 atom of trisilicate of soda + 3 atoms of trisilicate of lime + 12 atoms of silicate of alumina, and ought to behave before the blowpipe like paranthine or mesolite, which have very nearly the same composition. But this mineral so fully presents all the characters of felspar, in regard to fusibility and solubility with the fluxes, that it is difficult to imagine it not to be one. May not the mineral which Klaproth analysed be a compact, opaline scapolite?

Remark 2d.—Amongst the crystalline minerals that accompany meionite and nepheline, Werner has distinguished one by the name of Eisspat. This fossil behaves in every respect like felspar, and if, as Peschier has asserted, it contain soda, it follows that it is identical with the albite, or *kieselspath*, of Hausman.

8. *Elæolite*, Fettstein, from Fredrichsvarn, in Norway.[1]

Alone, in the matrass, gives off a little water, without any change in aspect or transparence.

On charcoal fuses pretty readily, with slight intumescence, into a blebby, colourless glass.

With borax dissolves easily, except a certain semi-transparent portion, which, as with mesotype, does not fuse at first with the rest, but requires a long continued blast.

With salt of phosphorus decomposes with the utmost difficulty, and leaves a siliceous skeleton. The glass becomes opaline on cooling.

With soda, vitrification extremely laborious. The glass is very difficult to fuse and obtain clear.

With solution of cobalt the fused edges are coloured blue.

9. *Andalusite,* from Fahlun. Felspath apyre.

Alone becomes covered with white spots (the

---

[1] Klaproth's analysis gives the very improbable composition of 1 atom of trisilicate of potassa + 4 atoms of silicate of alumina. Vauquelin found both soda and potassa in it. In point of composition, this mineral appears to be a scapolite, in which the lime is replaced by alcaline bases. B.

rest preserving its colour), and does not fuse either in thin spangle or in powder.

With borax fuses with difficulty, even in powder, into a transparent, colourless glass.

With salt of phosphorus decomposes with difficulty, and almost solely on the edges. The transparent part of the glass is not opaline.

With soda intumesces and decomposes, but does not fuse. The soda penetrates the charcoal, and a white mass remains on the surface.

With boracic acid and iron gives no phosphuret of iron.

10. *Apophyllite,* ichthyophthalmite, from Utö, and other places.[1]

Alone, in the matrass, gives off much water, and turns milk white.

On charcoal splits and dilates in the direction of the laminæ. In a strong heat, swells up like borax, and fuses, with continual intumescence, into a colourless, blebby glass.

With borax fuses easily into a transparent glass. The saturated glass becomes opaque by *flaming.*

With salt of phosphorus decomposes easily, and gives a silica skeleton, which usually intumesces so much as to fill the whole globule.

With soda fuses readily into a transparent glass, which, with a larger quantity of soda, becomes opaque on cooling.

[1] One atom of sexsilicate of potassa + 8 atoms of trisilicate of lime + 16 atoms of water. The formula calculated from the author's own analysis. C.

11. *Haüyne,* from Italy.

Alone, in the matrass, gives off no water.

On charcoal loses its colour, and fuses into a blebby glass.

With borax fuses, with effervescence, into a diaphanous glass, which becomes yellow on cooling, like the glass of gypsum. The saturated glass becomes opaque on cooling.

With salt of phosphorus fuses with effervescence, and leaves a silica skeleton; the glass is opaline on cooling.

Soda attacks it with difficulty, and converts it into a scoria, which is vitreous only at the extremity of the most projecting edge, and, on cooling, has the red colour of a hepar.

Remark.—These effects are so much like those exhibited by lapis lazuli, that we may presume the two minerals are very much alike in composition.

12. *Tourmaline,* Schörl (potassa-tourmaline).

(*a*). *Black, from Karingbricka.*

Alone, in the matrass, gives off no water.

On charcoal fuses with very brisk intumescence, and whitens. The intumesced portion fuses with difficulty into a semi-transparent yellowish grey globule.

With borax fuses easily, with some effervescence, into a transparent glass, which, when cold, has a slight tinge of iron.

With salt of phosphorus decomposes readily, with brisk effervescence, and leaves a skeleton of silica; the glass globule becomes opaline.

With soda dissolves, with much labour, into a difficultly fusible glass, which a further quantity of soda makes still more infusible.

(*b*). *Black, from Bovey,* in England.

Alone intumesces, and gives a black scoriaceous mass, of difficult fusion. Behaves with the fluxes like the preceding.

(*c*). *Green, from Brazil.*

Alone intumesces, blackens, vitrifies without fusing perfectly, and, after a good blast, gives a rounded, yellowish and blebby scoria.

With borax fuses pretty easily, with, at first, a slight effervescence. The glass has a slight tint of iron, and holds white particles in suspension, which, for a long while, resist solution.

With salt of phosphorus behaves like the preceding.

With soda like the preceding, but the glass is rather more fusible. On the platina foil no traces of manganese.

13. *Mica.* We have here another crystalline form, common to a multitude of different compounds, which often behave very dissimilarly before the blowpipe. Nevertheless, mica, as well as amphibole, garnet, &c., is susceptible of a certain general formula, in which it is only necessary to replace one isomorphous base by another, to deduce those peculiar to different varieties. But this chemical formula is not yet exactly known, and we do not well understand to what class of isomor-

phous bodies the combinations it includes belong. Trisilicate of potassa, combined with several atoms of silicate of alumina, is an essential part of mica; but with this, silicate of protoxide of iron is always combined; often also, silicate of protoxide of manganese, and sometimes silicate of peroxide of manganese.

We are ignorant how these substances are substituted for one another. The principal analyses of mica hitherto published, are, besides those of Klaproth and Vauquelin, the analyses lately made by M. H. Rose, in which he found that all mica contains more or less evident traces of fluoric acid, and a small quantity of water. But even these results are not capable of giving us a clear idea of the chemical formula representing mica. Rose's analyses indeed would give it, if in the silicates of iron and manganese the base were a peroxide; it would then be 1 atom of trisilicate of potassa + 12 atoms of silicate of alumina, in which we may replace a greater or less number of atoms of silicate of alumina, by a silicate of peroxide of iron or manganese. But Rose's experiments constantly proved that ferruginous mica, heated to redness in a retort, developed a green colour and acted on the magnet, without any disengagement of gas indicating the disoxidation of the presumed peroxide.

To these uncertainties are added the differences respecting the polarization of light by different varieties of mica, and the remarkable instance ad-duced by M. Biot, of a very magnesian mica which has but one axis, whereas the common micas have two.

As to the phenomena which the different sorts of mica present at a very high temperature, M. Rose observed that those which contain from $\frac{1}{4}$ to 1 per cent. of fluoric acid, lose their brilliancy, and become dull by ignition in close vessels. The others indeed lose their transparence, but they assume a semi-metallic, silvery, or golden lustre. The cause of the tarnishing of the micas that contain most fluoric acid, is evidently owing to the loss which the surfaces of the laminæ sustain by the fluoric acid carrying off with it a portion of their silica, in the form of silicated fluoric acid. It follows from these general considerations, that the micas, like the garnets, must vary with their localities, and that it is impossible, from experiments on them by the blowpipe, to deduce any distinctive character for the species, that may be common to all the varieties. I, therefore, proceed to detail the phenomena peculiar to some of them.

(a). *Mica,* from Broddbo and Finbo.[1] Occurs in granite.

Alone, in the matrass, gives off water at the melting point of glass, which contains decided traces

---

[1] One atom of trisilicate of potassa + 1 atom of trisilicate of protoxide of iron + 10 atoms of silicate of alumina. The formula calculated from Rose's analysis.* C.

---

* It contains besides 1·12 per cent. of fluoric acid. B.

of fluoric acid. At this heat the mica becomes dark green, rough to the touch, and dull on the surface. Heated to redness in the flame, by the blowpipe, it becomes white, or grey-white, and retains its brilliancy, but is covered with inequalities from the intumescence of the substance. It splits on the edges in the direction of the laminæ, and fuses into a blebby, yellowish grey glass.

With borax fuses readily, with effervescence, into a glass coloured green by iron. If previously heated to redness in the matrass, the fusion is not attended with effervescence.

With salt of phosphorus decomposes easily, intumesces so as to form a transparent skeleton, hardly perceptible, except by its disturbing the roundness of the globule. If only a small quantity of mica have been taken, the skeleton dissolves wholly in a good blast; but, with a larger quantity, the greater part becomes insoluble. The glass globule is opaline on cooling.

With soda swells up, and is converted into a tumified scoria, at first green, then grey, of which, only that side immediately exposed to the flame, fuses into a transparent, slightly green glass. On the platina foil gives very decided traces of manganese.

With solution of cobalt gives a black glass.

(b). *Mica,* from North America. Occurs in granite.

Alone, in the matrass, gives off water without becoming opaque. At the fusing point of glass, it assumes a silvery metallic whiteness; gives scarcely perceptible traces of fluoric acid.

On charcoal turns milk white, and then, at a very high temperature, fuses on the extreme edge into a white enamel; the thinnest spangle cannot be fused into a globule.

With borax fuses at first with slight effervescence, afterwards quietly. If previously heated red, till it becomes opaque, it fuses without any effervescence with borax.

With salt of phosphorus in small quantity, fuses at first completely. Afterwards decomposes with difficulty, and leaves a very small silica skeleton. The glass becomes opaline, but only after a prolonged blast.

With soda a white scoria, that may be fused into a transparent glass on the points most exposed to the flame.

With solution of cobalt fused edges coloured blue.

(c). *Mica, from the limestone quarry at Pargas.* The assay was part of a hexahedral prism.

Alone, in the matrass, gives off a little water, without change of aspect. In a strong heat gives no trace of fluoric acid. Preserves its smoky colour and transparence even when heated red by the blowpipe flame. Fuses easily into a milk-white glass, which may be obtained in a globule. The unfused portion of the assay loses nothing of its transparence.

With borax dissolves readily without the slightest effervescence. The lamina of mica lies quietly in the flux, and, retaining its transparence, diminishes by degrees till it entirely disappears.

With salt of phosphorus decomposes readily, and leaves a perfectly transparent silica skeleton. The glass is opaline.

With soda intumesces, turns milk-white, and then fuses into an opaque globule, which becomes milk-white on cooling.

With solution of cobalt gives a clear blue on the fused edges.

Remark.—Did not the exterior form of the three last minerals denote that they are micas, it would not be natural to refer them to the same species, from the phenomena they present with the blowpipe. The fusibility of the mica from Pargas, and the difficulty, not to say impossibility, of fusing that from America, are indications of a very different chemical composition. These differences correspond with those we have observed between common schörl and the tourmaline apyre, whose composition is also very dissimilar.

14. *Talc.* What I have said of the micas, applies very well to the talcs, as the following experiments will show:—

(*a*). *Clear-green transparent talc*, from the valley of Bine.

Alone gives off no water, nor loses its transparence by ignition. In a strong heat it exfoliates, whitens on the part most heated, but does not fuse.[1]

---

[1] "All the foliated varieties of this mineral are fusible into a *greenish glass*." *Clarke.* Gas Blowpipe, p. 54.

With borax fuses easily, and with brisk effervescence, into a transparent glass.

With salt of phosphorus fuses easily, with effervescence, into an opaline glass, and leaves a transparent siliceous skeleton.

With soda intumesces, and changes into a white, semi-fused scoria.

With solution of cobalt gives a very pale red colour.

(*b*). *White opaque talc*, from the valley of Fenestrolle.

Behaves like the preceding, except that it effervesces less with the fluxes.

(*c*). *Translucid greenish talc*, from Skyttgrufwa, near Fahlun.

Alone, in the matrass, no trace of water, clears a little on the edges, but does not lose its transparence by a moderate ignition. In a stronger heat it whitens, becomes scaly and rounded on the edge into a white, blebby mass.

With borax fuses, with effervescence, into a transparent glass, which, whilst hot, is tinged with iron. A portion resists solution at first, but afterwards fuses very slowly without effervescence.

With salt of phosphorus decomposes easily; gives an almost transparent skeleton of silica, and an opaline glass.

With soda intumesces, and fuses into an opaque, difficultly liquefiable glass, which may be obtained clear with a certain proportion of soda. In general, this glass absorbs a large quantity of it.

With solution of cobalt gives a red colour, with a strong heat.

Remark.—Thus the species of talc found in the mine at Fahlun behave before the blowpipe.

(*d*). *White talc,* from China; Agalmatolite.

Alone, in the matrass, gives off an empyreumatic water. The assay blackens, as commonly happens with the serpentines, and silicates of magnesia.

On charcoal whitens with heat, becomes scaly on the surface, and presents some marks of fusion at the extremity of the most projecting part.

With borax behaves like the preceding minerals, except that the glass is colourless.

With salt of phosphorus no decomposition. At first a brisk effervescence ensues, and the assay diminishes in bulk; the remainder appears absolutely insoluble.

With soda, and the solution of cobalt, behaves like the preceding mineral.

(*e*). *Black talc,* from Finbo, near Fahlun.

Alone, in the matrass, gives off a large quantity of water, which, by igniting the assay, presents evident traces of fluoric acid. At a red heat, the mineral assumes a clearer colour and a greenish tint; it afterwards fuses pretty easily into a black glass.

With borax fuses easily, without remarkable effervescence, into a glass strongly coloured by iron.

With salt of phosphorus decomposes easily, and leaves a skeleton of silica. Whilst hot, the glass is tinged by iron; it becomes opaline on cooling.

With a little soda dissolves into a black glass, which a larger dose of soda makes difficult to fuse, and colours yellowish brown. On platina foil, it presents very slight traces of manganese.

Amongst the fossils, considered as the remains of an ancient organization, there are but two whose nature can be ascertained by the blowpipe. They are both aluminous salts.

1. *Ammoniacal alum*, from Tschermig, in Bohemia.

Alone, in the matrass, gives off water and swells up. A sublimate of sulphite of ammonia then forms, the greater part of which is dissolved by the water, and sulphurous acid is disengaged. After ignition of the mass, the remainder behaves like pure alumina. Kneaded with *soda*, and gently heated on platina foil, it exhales a very sensible odour of ammonia.

2. *Mellite.* Mellate of alumina.

Alone, in the matrass, gives off water, whitens, and becomes opaque. At a red heat, carbonizes without exhaling any very sensible empyreumatic odour; the water condensed in the matrass is colourless, and has no acid or alcaline action.

On charcoal first blackens, takes fire, then in an intense heat becomes white, and contracts considerably; it afterwards retains its whiteness, and behaves like pure alumina.

Remark.—Mellate of iron has been found lately, but I have not yet been able to procure a specimen of it for examination with the blowpipe.

MINERALS NOT HITHERTO ANALYSED, AND WHOSE RANK, THEREFORE, IS NOT YET FIXED IN THE CHEMICAL ARRANGEMENT.

1. *Helvine*, from Schwartzenberg, in the metalliferous mountains of Saxony. Furnished by M. Cordier.

Alone, in the matrass, gives off water, without its sulphur-yellow colour, or transparence, being altered.

On charcoal fuses with bubbling, in the interior flame, into an opaque globule, nearly of the same colour as the stone. In the exterior flame, the fusion is much more difficult, and the colour of the mineral turns to brown.

With borax fuses slowly into a diaphanous glass, which, as long as any portion of the assay remains undissolved, is yellowish, and retains a yellow tint even on cooling. After complete solution, the glass becomes colourless in the interior flame, and deep amethyst colour in the exterior; the tint in the latter case is not perfectly pure.

With salt of phosphorus decomposes easily, and gives a siliceous residuum, and a glass colourless both cold and hot, but which becomes opaline on cooling.

With soda at first intumesces, then fuses easily into a black glass globule, which, in the reducing flame, becomes chesnut-brown. *On platina foil,* swells up, divides, and becomes chesnut-brown, without colouring the soda; but, with a prolonged blast, the whole mass assumes the green colour of manganese, and is converted into mineral cameleon.

Remark.—This mineral is, therefore, a silicate of manganese, into which that metal enters as an essential constituent; whether it contain any other base, as lime or alumina, its reactions do not inform us. It appears, however, that iron is not one of its essential elements.

2. *Chiastolite, Macle,* from Britanny.

Alone gives off a little water, without changing its aspect; whitens by heat, but does not fuse. A very thin cake, made with the pulverised mineral, concretes into a mass.

With borax fuses with extreme difficulty, even in powder, into a transparent glass.

Salt of phosphorus has scarcely any action on this mineral, but it becomes colourless and transparent in the globule. If the pulverised assay be added in very small doses, it dissolves in the flux without residuum; but the salt of phosphorus soon becomes saturated, and refuses to dissolve a further quantity.

With soda the assay decomposes and swells up, but neither fuses nor forms a scoria.

With solution of cobalt we obtain a blue colour, whose purity is proportionate to that of the assay.

Remark.—These results show that chiastolite is a silicate of alumina, and to all appearance a subsilicate.

PHENOMENA

DEVELOPED BY URINARY CALCULI BEFORE THE BLOWPIPE.

It is very important to medical men to know the nature of the concretions formed in the urinary passages of the patient who applies to them for relief. The chemical composition of these substances is more easily ascertained than is generally supposed, and the blowpipe furnishes us, for that purpose, with a method of proving them, as simple as it is infallible, and which requires no more chemical knowledge than every physician ought to possess.

1. *Urinary calculi, formed of uric acid.*

Heated by themselves on charcoal, or platina foil, they carbonise, fume, and develope an animal odour; in the exterior flame, they continually diminish in bulk. Towards the end of the roasting, they burn with increased light; although we then suspend the blast, the matter still continues to burn brilliantly, and at last leaves a residuum, consisting of a very small quantity of strongly alcaline white ash.

As there are other combustible substances that might be confounded with uric acid, a portion of the calculus should be examined in the moist way, as follows: We place 1-10th of a grain of the substance on a thin lamina of platina or glass, and having added one drop of nitric acid, heat the whole over the lamp; the uric acid dissolves with effervescence, and the mass must then be carefully dried, lest it burn; when dry, a beautiful red colour appears. If the assay contain but a small quantity of uric acid, it sometimes blackens, instead of reddening by the heat. In that case we must take a fresh portion, and having dissolved it in nitric acid, withdraw it from the heat, when the solution is nearly dry, and then leave it to cool till the desiccation is completed. Then, turning the support to which the assay adheres, upside down, we hold it in that position above a little caustic ammonia, placed over the lamp; as soon as the ammoniacal vapour reaches the dried substance, the fine red colour is developed.[1] The same colour is developed, but not so fine, if we moisten the dried matter with a little weak solution of ammonia.

Urinary calculi, formed by a mixture of uric acid and the earthy phosphates, are sometimes met with. They carbonize and burn away like the first, but leave a larger residuum, which is neither alcaline nor soluble in water. Treated with nitric acid and ammonia, they exhibit the beautiful red colour which distinguishes uric acid. The remaining ash is either phosphate of lime, or phosphate of magnesia, or a mixture of the two.

[1] This mode of applying the ammonia was invented by Professor Jacobsen of Copenhagen, who, by this reagent, has shown the existence of uric acid in the excretions of animals of the lowest order. B.

2. *Calculi formed of urate of soda.*

This substance rarely forms part of urinary calculi, and is seldom met with except in the hard excrescences that form round the articulations of gouty patients. *Alone, on charcoal,* they blacken, develope an empyreumatic animal odour, are difficultly reduced to ashes, and leave a strongly alcaline, grey residuum, which forms a glass by fusion with a little silica. If, as is most common, the calculus contain earthy salts, the glass is white, or greyish white, and opaque.

3. *Calculi formed of urate of ammonia.*

Behave before the blowpipe like uric acid calculi. Treated with a drop of caustic potassa, they exhale, when gently heated, a strong odour of ammonia. The slight ammoniacal odour, which potassa developes with almost all animal substances, has nothing to do with this. These calculi often contain, besides, urate of soda.

4. *Urinary calculi of phosphate of lime.*

Alone, on charcoal, they blacken, exhale an animal empyreumatic odour, and finally become white; they do not fuse; in other respects they behave like phosphate of lime. (See *phosphate of lime,* p. 251.) A proof that these calculi are not siliceous, is that they swell up with soda without vitrifying, and when dissolved in boracic acid, and then fused with a little iron, they give a regulus of phosphuret of iron.

5. *Calculi of ammoniaco-magnesian phosphate.*

Alone, on the platina foil, they exhale a strong odour of salt of hartshorn, blacken, swell up, and lastly become greyish white. They fuse easily into a greyish white, enamel-like globule.

With borax and salt of phosphorus they fuse into a transparent glass, which, if the proportion of the assay be large, becomes milk white on cooling.

With soda they fuse into an intumescent white scoria, which a larger dose of soda renders infusible.

With boracic acid and iron they readily give a regulus of phosphuret of iron.

With nitrate of cobalt they give a dark red glass.

When the calcareous and ammoniaco-magnesian salts occur together, it is known by the diminished fusibility of the mixture [1].

6. *Calculi of oxalate of lime.*

Alone, immediately exhale an urinous odour. Those whose crystallization is least confused, lose

---

[1] This is a mistake,—and I am obliged to my friend Dr. Marcet for having pointed it out. The compound calculus is *much more fusible* than either of its component salts—it has in consequence been named, κατ'ιξοχην, " *the fusible calculus.*" See Marcet's Essay on Calculous Disorders, p. 78. C.

the polish of their surface, whilst their colour gets clearer. After a moderate ignition the residuum effervesces with a drop of nitric acid, and, with a good heat, leaves quick lime on the charcoal, which acts on reddened litmus paper like an alkali, and commonly falls to powder when slacked with water, but not if the residuum contain phosphate of lime.

7. *Siliceous calculi.*

Alone, leave an infusible, sometimes scoriaceous ash, wnich fuses with a small quantity of soda, slowly and with effervescence, into a more or less transparent glass globule.

8. *Cystic oxide calculi.*

These calculi behave very nearly like those of uric acid, before the blowpipe; they do not fuse, they take fire readily and burn with a bluish green flame, exhaling a very acid, peculiar odour, which has some remote resemblance to that of cyanogen. Their ash is not alcaline, and in a good heat fuses into a greyish white mass. They differ from uric acid, both by the odour they develope by heat, and by not producing the red colour by the action of nitric acid.

Remark.—I have not had an opportunity of examining the calculi discovered by Dr. Marcet, in which he found a peculiar substance that he has named *xanthic oxide*.[1]

---

[1] The following account of *xanthic oxide* is taken, by his permission, from Dr. Marcet's " Essay on the Chemical

History, and Medical Treatment of Calculous Disorders," to which excellent work I refer the reader, who wishes to obtain a perfect knowledge of calculi in general. The description of the xanthic oxide is most accurate, as I can testify from my own observation, Dr. Marcet having had the goodness to allow me to inspect the calculus itself, and even to sacrifice a further portion of it, on purpose to show me its distinguishing chemical properties.

" 1. It was, when entire, of an oblong spheroidal shape, and weighed only about eight grains.

" 2. Its texture is compact, hard, and laminated, surface smooth. It is of a reddish cinnamon colour, which is much heightened on adding caustic alkali to the calculus in powder. Between the red laminæ, faint whitish lines are perceived.

" 3. When the blowpipe is applied, it crackles, splits into small pieces, turns black, and is ultimately consumed, leaving only a minute particle of white ash. The smell it emits is that of an animal substance, and is peculiar, though feeble and not easily defined. It does not at all resemble that of the lithic (uric) acid, or of the cyanic oxide.

" 4. When exposed to destructive distillation, it crackles, splits into scaly fragments, blackens, and emits a fœtid ammoniacal liquor, from which carbonate of ammonia crystallizes on cooling, and a heavy yellowish oil.

" 5. When reduced to an impalpable powder by scraping, and boiled in water, the greater part of it is dissolved, and this solution slightly reddens litmus paper. If the clear liquor be decanted off, and allowed to cool, it covers itself with a white flocculent film, apparently not crystalline, which gradually subsides, forming a white incrustation ; and if the glass be scratched with a pointed instrument, during or just before this deposition, white lines appear at the points of contact, as in the case of the ammoniaco-magnesian phosphate.

" 6. Caustic potash dissolves this calculus very readily, and it may be precipitated from this solution by acetic acid, provided the latter be not added in great excess. It is also soluble in ammonia, and in the alkaline subcarbonates.

" 7. The mineral acids also dissolve it, though not near so

readily as the alkalies; so that a doubt may arise whether acids may not act upon it merely through the water they contain.

"8. The residues of its solution in the muriatic and sulphuric acids are white; and, as far as I was able to judge from the minute quantities of the calculus which I was able to spare for experiments, no distinct crystals were formed. Concentrated sulphuric acid does not blacken this calculus.

"9. When the solution of the new substance in nitric acid is evaporated to dryness, the residue assumes a bright lemon colour. This yellow residue is partly soluble in water, to which it communicates its colour. The addition of an acid takes away this yellowness; but if caustic potash be added to the yellow substance, it instantly turns it to a more or less intense red colour, according to the degree of dilution; and upon evaporation it assumes a brilliant crimson hue, which, however, disappears on adding water, the yellow colour being reproduced, and remaining perfectly transparent. The previous action of nitric acid is necessary for these singular changes; for if the potash be added to the pure calculous substance, such as deposited by water, no change of colour takes place. The residue of the solution of the calculus in water produces the yellow substance, when treated with nitric acid, just the same as the calculus itself.

"10. The new substance is insoluble in alcohol or ether.

"11. It is but very sparingly soluble in acetic acid.

"12. It is insoluble, or nearly so, in oxalic acid.

"13. It appears to be insoluble, or nearly so, in bicarbonate of potash, or saturated carbonate of ammonia.

"Upon the whole, this calculus appears to be a substance *sui generis*, and will probably be found entitled to be considered as an oxide, though it is certainly much less soluble in acids than the cystic oxide.

"It is considerably more soluble in water than lithic acid, and is abundantly distinguishable from it by the lemon colour it forms when acted upon by nitric acid, and by its smell when burnt.

"It is as easily distinguished from the cystic oxide, since

the latter forms a white residue on evaporating its nitric solution; has a smell quite peculiar to itself; is not formed of laminæ; and is rather more soluble in alkalies, and much more soluble in acids than the new substance in question. Should there remain any doubt respecting the peculiar nature of this calculus, I would beg to add that both Dr. Wollaston and Dr. Prout have examined its leading properties, and have expressed their conviction that it could not be referred to any of the species hitherto described.

"It is so difficult to find an appropriate and unexceptionable denomination for a new substance, that it is with reluctance I propose one for this, especially as a similar production may possibly never again be noticed. It has occurred to me to name it *xanthic* calculus, from ξανθός, yellow, because this term alludes to a striking and probably characteristic property of the substance in question, (that of forming a lemon-coloured compound when acted upon by nitric acid), and yet does not imply any systematic notion as to its composition." C.

INDEX.

A.

Acid, fluoric, 127
—— hydriodic, 126
—— muriatic, 125
—— nitric, 125
—— phosphoric, 128
—— sulphuric, 125
Actinote, asbestiform, 263
Alaunstein, 310
Alalite, 268
Albite, 307
Alcalies, 74
—— how distinguished, note, 76
Allanite, 291
Allochroite, 284
Allophane, 222
Almandine, 224
Alum, 310
—— ammoniacal, 325
Alumina, 81
—— how discovered by nitrate of cobalt, 60
Aluminite, 214
Amalgam, native, 149
Amblygonite, 297
Ammonia, 76
Amphibole, 258
Amphiboles, aluminous, 264
—— colourless, 261
—— non aluminous, 261
Amphigene, 311
Analcime, 307
Anatase, 140
Andalusite, 314
Anthophyllite, 289

Antimony, 90
—— alloys of, 123
—— alloy with arsenic, 139
—— how distinguished from bismuth and tellurium, 111
—— native, 137
—— oxide of, 140
—— sulphuret, 137
Antimonic acid, 91
Antimonious acid, 91, 140
Anvil, described, 39
Aplome, 279
Apophyllite, 315
Arragonite, 250
Arsenic, 135
—— alloys of, 121
—— how detected; 121
—— scapiform, 180
—— sulphuret, red, 135
————— white, 136
————— yellow, 135
Arseniates, 133
Arsenious acid, sublimate of, how distinguished from oxide of tellurium, 124
Asbestus, 263
Axinite, 288.

B.

Baryta, 76
—— carbonate of, 296
—— sulphate of, 295
Beryl, 233

Bismuth and tellurium, alloys of, 151
——— and selenium, 151
——— how distinguished from antimony and tellurium, 111
——— native, 150
——— oxide of, 153
——— sulphuret, 150
Bitter salt, 235
Bitter spar, 251
Blattererz, 156
Blowpipe, Bergman's, 9
——— blast of the, 20
——— Brooke's, note, 15
——— description of, 8
——— fixed one, note, 14
——— Gahn's, 9
——— history of, 1
——— Pepys's, note, 9
——— Tennant's, 10
——— Wollaston's, 11
Bone ashes, how used in cupellation by the blowpipe, 64
Boracic acid, 130, 135
——— (vitrified) its use as a reagent, 59
Boracite, 236
Borax, 302
——— its use as a reagent, 54
Botryolite, 252
Bournonite, 155
Brazil wood paper, effect of fluoric, phosphoric, and oxalic acids on, note, 127
Byssolite, 264.

C.

Calaite, 216
Calculi, ammoniaco-magnesian phosphate, 331
——— cystic oxide, 332
——— oxalate of lime, 331
——— phosphate of lime, 330
——— siliceous, 332
Calculi, urate of ammonia, 330
——— urate of soda, 330
——— uric acid, 328
——— xanthic oxide, 332
Carbonic acid, 130
Carpholite, 222
Cerine, 290
Cerite, 211
Cerium, fluate of, 209
Chabasic, 277
Charcoal, of alder recommended, note, 32
——— of pine the best, 30
——— used as a support, 30
Chiastolite, 327
Chlorite, 243
Chondrodite, 237
Chromates, 133
Chrome, earthy, 136
Chrysoberyl, 278
Cinnabar, 144
——— mealy, 144
Clay, Cologne, 228
——— Stourbridge, 228
Cobalt, arseniate of, 182
——— arsenical, 180
——— arsenite of, 182
——— black oxide of, 181
——— nitrate of, its use as a reagent, 60
——— sulphuret of, 179
Columbite, 202
Combustibles, of the, 18
Copper and antimony, sulphuret of, 163
——— argentiferous sulphuret of, 163
——— arseniate of, 171
——— carbonate of, 170
——— oxides of, 168
——— oxide of, its use as a reagent, 65
——— phosphate of, 169
——— pyrites, 164
——— seleniuret of, 167
——— submuriate of, 169
——— sulphate of, 168
——— sulphuret of, 162
Cryolite, 302

Cymophane, 278
Cyphrine, 286.

D.

Datholite, 252
Davy, his views respecting flame, note, 23
Decrepitation, how guarded against, 36
Diallage, 243
Diaspore, 226
Dichroite, 242
Diopside, 268
Dipyre, 276
Disthene, 219
——— employed as a support, 35
Dunkel, Weissguttigerz, 156.

E.

Ecume de mer, 239
Egerane, 286
Eisensinter, 193
Ekebergite, 308
Electrum, 149
Elæolite, 314
Emerald, 233
Endellione, 155
Epidote, 272
——— manganesian, 273
Essonite, 284
Euchairite, 167
Euclase, 234.

F.

Fahlerz, ores, 168
Fahlunite, 221
——— hard, 241
Feldspar, 312
Flame, Davy, his views respecting, note, 23

Flame, directions concerning, 21
——— Sym on the structure of, note, 27
Flaming, meaning of the term, 55
Fluate, easy mode of ascertaining a, note, 128
Fluoric acid, 127
Fluor spar, its use as a reagent, 59
Forceps described, 37
——— Popys's, for decrepitating substances, note, 38
Fuller's earth, 227.

G.

Gadolinite, 230
Gahnite, 187
Garnet, 225
——— black, 282
——— clear brown, 282
——— deep brown, 283
——— green, 282
——— red, 283
Garnets, magnesian, 245
Gehlinite, 289
General rules for experiments with the blowpipe, 65
Glanzerz, 146
Glass tubes, how used in roasting assays, 35
Glauberite, 301
Glucina, 82
Gold, 118
——— graphic, 142
——— telluriferous and plumbiferous, 143
Grammatite, 262, 264
Graphite, 190
Gypsum, its use as a reagent, 59.

H.

Hammers, best form for, 39
Harmotome, 296
Haüyene, 316
Hedenbergite, 270
Helvine, 326
Hornblende, black, crystallized, 266
——————— in large laminæ, 266
——————— primitive, 265
Horn lead, 157
Hyacinth, 212
Hydrates, 130
Hydriodic acid, 126
Hyperstene, 244.

I.

Idocrase, 285
Iridium, 118
——— and osmium, alloy of, 141
Iron, arseniate of, 192
—— carbonate of, 192
—— chromiferous, 194
—— columbiferous, 190
—— hydrate of, 196
—— its use as a reagent, 63
—— magnetic ore of and peroxide of, 191
—— native, 188
—— phosphate of, 192
—— sulphate of, 191
—— sulphuret of, 188
—— titaniferous, 195.

J.

Jade, 241.

K.

Kieselmalachite, 173

Kobaltglanz, 181
Konpholite, 274
Kupfermangan, 208.

L.

Lamp, form of, note, 19
Lapis lazuli, 305
Laumonite, 277
Lazulite, 215
Lead, arseniate of, 158
—— carbonate of, 157
—— chloro-carbonate of, 157
—— chromate of, 160
—— its use as a reagent, 63
—— molybdate of, 159
—— oxide of, 156
—— phosphate of, 158
—— sulphate of, 156
—— sulpho-carbonate of, 157
—— sulphuret of, 153
—— tungstate of, 161
Lepidolite, 300
Licht Weissgüttigerz, 155
Lievrite, 278
Lime, 79
—— arseniate of, 253
—— carbonate of, 250
—— fluate of, 248
—— phosphate of, 251
—— sulphate of, 247
—— tungstate of, 254
—— uranate of, 255
Lithia, 75
Loböite, 286.

M.

Macle, 327
Magnesia, 81
——— how discovered by nitrate of cobalt, 60
——— hydrate of, 247
Magnesite, 235
Malacolite, white, 268

INDEX.

Manganese, carbonate of, 200
——————— ferriferous phosphate of, 199
——————— hydrate of, 208
——————— peroxide of, 198
——————— silicate of, 207
——————— sulphuret of, 196
Mangankiesel, black, 205
——————— red, 206
Matrass, use of, 35
Meionite, 312
Melanite, 282
Mellite, 325
Mercury, chloride of, 145
——————— hepatic, 145
Mesolite, 307
Mesotype, 306
Metallic carburets, 124
——————— seleniurets, 120
——————— sulphurets, 119
Mica, 317
—— used as a support, 35
Mispickel, 189
Molybdates, 133
Molybdena, sulphuret of, 137
Molybdic acid, 85, 137
Moroxite, 251
Muriatic acid, 125.

N.

Nadelerz, 166
Nepheline, 220
Nephrite, 241
Nickel, arseniate of, 177
——— arsenical, 175
——— spiesglanserz, 176
——— sulphuret of, 175
——— white ore of, 176
Nitrate of cobalt, its use as a reagent, 60
Nitric acid, 125.

O.

Ochre, vitriol, 191

Orthite, 292
Oxidation, how produced, 28
Oxide of antimony, 91
——— bismuth, 110
——— cadmium, 104
——— cerium, 100
——— chrome, 89
——— cobalt, 107
——— columbium, 93
——— copper, 115
——— iron, 105
——— lead, 114
——— manganese, 101
——— mercury, 117
——— nickel, 108
——— silver, 117
——— tellurium, 92
——— tin, 112
——— titanium, 94
——— uranium, 99
——— zinc, 103.

P.

Palladium, 118
——————— native, 145
Paranthine, 75
Pargasite, 267
Peridot, 242
Petalite, 298
Phosphoric acid, 128
Phosphorus, salt of, its use as a reagent, 56
Pimélite, 178
Pinite, 220
Pistasite, 273
Platina, 118
——— as a support, 32
——— foil, 33
——— grains of, 142
——— spoon, 32
——— wire, 33
Pleonaste, 246
Plombgomme, 162
Polyhalite, 309

Potassa, inferior as a reagent to soda, 54
Prehnite, 274
Pyenite, 218
Pyrallolite, 238
Pyrites, common sulphureous, 188
———— magnetical, 188
Pyrope, 287
Pyrorthite, 293
Pyrosmalite, 206
Pyroxene, 267.

R.

Reagents, 45
Reduction, how produced, 28
Rhodium, 118
Roasting, how performed, 35
Rubellite, 308
Rutilite, 140.

S.

Salite, pale green, 268
Salt of phosphorus, its use as a reagent, 56
Saltpetre, 311
———— its use as a reagent, 58
Scapolite, 275
Scherbenkobalt, 180
Scolezite, 276
Scorodite, 193
Sea salt, 302
Seifenstein, 240
Seleniates, 133
Seleniurets, metallic, 120
Serpentine, common, 240
———— noble, 240
Silica, 83, 141
———— antimonial, 148
———— brittle sulphuret of, 147
———— chloride of, 149
———— its use as a reagent, 65

Silica, red, 146
———— sulphuret of, 146
Silicates, 130
Soda, its use as a reagent, 45
———— sulphate of, 301
———— what metals reducible by, 52
Sodalite, 303
Sordawalite, 245
Speiskobalt, 180
Sphene, 256
Spiesglansbleyerz, 155
Spinel, 245
Spodumene, 298
Staurotide, 223
Stilbite, 277
Strontita, 78
———— carbonate of, 295
———— sulphate of, 294
Sulphur, how detected, 119
Sulphurets, metallic, 119
Sulphuric acid, 125
Supports, of the, 30
Sym on the structure of flame, note, 27.

T.

Tabular spar, 257
Talc, black, 324
———— clear green transparent, 322
———— translucid greenish, 323
———— white, from China, 324
———— white, opaque, 323
Telesia, 213
Tellurium, alloys of, 123
———— how distinguished from antimony and bismuth, 111
———— sublimed oxide of, how distinguished from arsenious acid, 124
Tin, its use as a reagent, 62
———— oxide of, 153
———— sulphuret of, 165

Titanite, 140
Topaz, 217
Tourmaline, black, 316
——————— from Bovey, 317
——————— clear blue, finely striated, 299
——————— dark blue, in large crystals, 299
——————— green, 317
——————— red, and clear green, 298
Tremolite, asbestiform, 262
Triphane, 298
Tungstates, 133
Tungstic acid, 87.

U.

Uranium, protoxide of, 183
——————— yellow hydrated oxide of, 183.

V.

Vauqueline, 172
Vesuvian, 285.

W.

Water, see hydrates, 130
Wavellite, 214
Weissgüttigerz, dunkel, 156
——————— licht, 155
Wolfram, 201
Wurfelerz, 193.

Y.

Yenite, 278
Yttria, 83
——— and cerium, fluate of, 229
Yttrocerite, 249
Yttrocolumbite, 229.

Z.

Zeolite, mealy, 277
Zinc, blende, 184
——— carbonate of, 185
——— oxide of, 184
——— silicate of, 186
——— subcarbonate of, 186
——— sulphate of, 185
Zircon, 212
Zirconia, 83
Zöizite, 272.

ERRATA.

P. 19, note, last line; for "*of the size drawn in the figure,*" read, "*about twice the size of the figure.*"
P. 32, l. 19, for "*soda or charcoal,*" read "*soda on charcoal.*"
P. 137, l. 18, for "*earthly,*" read "*earthy.*"
P. 140, last line, for "*give,*" read "*gives.*"
P. 141, l. 11, for "*qualities,*" read "*quantities.*"
P. 154, l. 15, for "*copper ores,*" read "*some copper ores.*"
P. 211, note, l. 3 from bottom, for "*silicate,*" read "*sub-sesqui-silicate.*"
P. 214, l. 23, for "*phosphoric acid,*" read "*salt of phosphorus.*"
P. 228, l. 6, for "*the fusion,*" read "*their fusion.*"
P. 248, note, l. 2 from bottom, for "4" read "2."
P. 259, l. 13, insert a comma after "these."
P. 298, l. 10 from bottom, after Utö, add "(Lithium Tourmaline)."

P. 75, l. 13,
P. 88, l. 21,
P. 94, l. 25,
P. 95, l. 1,
P. 97, l. 7,
P. 106, l. 3, } for "*dull,*" read "*dark.*"
P. 107, last line,
P. 108, l. 23 and 24,
P. 110, l. 5,
P. 112, l. 7,
P. 147, l. 16,

THE END.

C. Baldwin, Printer,
New Bridge-Street, London.

Titanite, 140
Topaz, 217
Tourmaline, black, 316
——————— from Bovey, 317
——————— clear blue, finely striated, 299
——————— dark blue, in large crystals, 299
——————— green, 317
——————— red, and clear green, 298
Tremolite, asbestiform, 262
Triphane, 298
Tungstates, 133
Tungstic acid, 87.

U.

Uranium, protoxide of, 183
——————— yellow hydrated oxide of, 183.

V.

Vauqueline, 172
Vesuvian, 285.

W.

Water, see hydrates, 130
Wavellite, 214
Weissgüttigerz, dunkel, 156
——————— licht, 155
Wolfram, 201
Wurfelerz, 193.

Y.

Yenite, 278
Yttria, 83
——— and cerium, fluate of, 229
Yttrocerite, 249
Yttrocolumbite, 229.

Z.

Zeolite, mealy, 277
Zinc, blende, 184
——— carbonate of, 185
——— oxide of, 184
——— silicate of, 186
——— subcarbonate of, 186
——— sulphate of, 185
Zircon, 212
Zirconia, 83
Zöizite, 272.

Lightning Source UK Ltd.
Milton Keynes UK
UKHW050311240119
335965UK00011BA/1084/P